Patrick Guidotti
Advanced Mathematics

Also of Interest

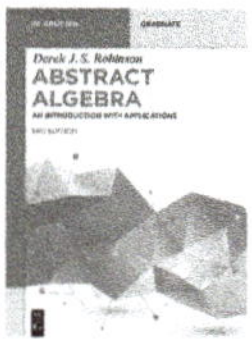

Abstract Algebra
An Introduction with Applications
Derek J. S. Robinson, 2022
ISBN 978-3-11-068610-4, e-ISBN (PDF) 978-3-11-069116-0

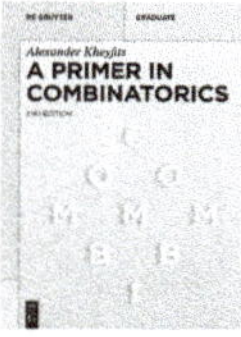

A Primer in Combinatorics
Alexander Kheyfits, 2021
ISBN 978-3-11-075117-8, e-ISBN (PDF) 978-3-11-075118-5

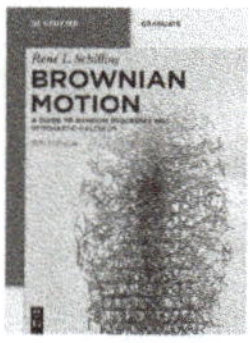

Brownian Motion
A Guide to Random Processes and Stochastic Calculus
René L. Schilling, 2021
ISBN 978-3-11-074125-4, e-ISBN (PDF) 978-3-11-074127-8

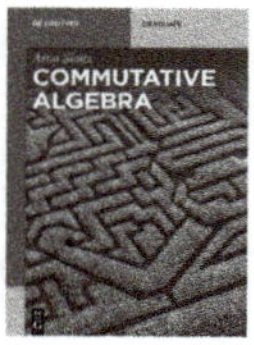

Commutative Algebra
Aron Simis, 2020
ISBN 978-3-11-061697-2, e-ISBN (PDF) 978-3-11-061698-9

General Topology
An Introduction
Tom Richmond, 2020
ISBN 978-3-11-068656-2, e-ISBN (PDF) 978-3-11-068657-9

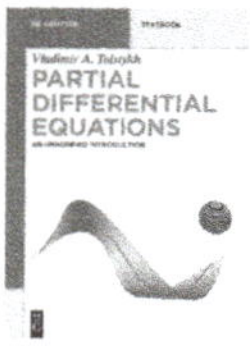

Partial Differential Equations
An Unhurried Introduction
Vladimir A. Tolstykh, 2020
ISBN 978-3-11-067724-9, e-ISBN (PDF) 978-3-11-067725-6

Patrick Guidotti

Advanced Mathematics

An Invitation in Preparation for Graduate School

DE GRUYTER

Mathematics Subject Classification 2010
15-01, 26-01, 34-01, 35-01, 40-01, 46-01, 47-01, 49-01, 53-01, 54-01, 57-01, 60-01, 97-01, 97F40, 97F50, 97H60, 97I20, 97I30, 97I40, 97K50, 97K60

Author
Prof. Patrick Guidotti
University of California
Department of Mathematics
340 Rowland Hall
Irvine CA 92697
USA
gpatrick@math.uci.edu

ISBN 978-3-11-078085-7
e-ISBN (PDF) 978-3-11-078092-5
e-ISBN (EPUB) 978-3-11-078098-7

Library of Congress Control Number: 2022938213

Bibliographic information published by the Deutsche Nationalbibliothek
The Deutsche Nationalbibliothek lists this publication in the Deutsche Nationalbibliografie; detailed bibliographic data are available on the Internet at http://dnb.dnb.de.

Cover image: Patrick Guidotti, Alpi Ticinesi
Typesetting: VTeX UAB, Lithuania
Printing and binding: CPI books GmbH, Leck

www.degruyter.com

To the women of my life: Irene, Stelia, Natalia, Alessia, Lara, and Milena

Follow your intuition but steadily question it.
Learning often amounts to dispelling one's preconceived notions.

Contents

Foreword

Equivocation is ubiquitous in human language. In some situations, it is harmless, sometimes even funny, like when the realization of a misunderstanding creates an awkward moment. In other situations, it can be used for manipulation, like in politics and in advertising. It is an unavoidable feature of any language because we humans tend to read and interpret words and sentences not simply in light of meanings found in a dictionary but also considering the source, the circumstances, and our own prior experiences. We sometimes do not hear what is said to us because we have filters on, whereas other times we hear what we want to hear and may not have been said. Mathematics, as a language that distinguishes itself from all other human languages, tries to remove ambiguity as much as possible. It does so by providing precise definitions and by requiring a standard of proof that any argument must satisfy before it is accepted as valid. While a mathematician is completely free to define the objects of her study and is therefore not in the least limited by the rigor of the discipline, once these objects have been identified and described, one has to abide by the rules they entail. It is my experience as a teacher that the source of many a student's difficulties with the subject can often be traced back to one of two phenomena. One consists of an understanding that is purely *intuitive*, where words (mathematical symbols and terminology) evoke approximative, sometimes even precise meaning, but still do not lead to the targeted ability to perform the mathematical manipulations necessary to derive further understanding and knowledge or even the desirable ability to apply the understanding to any concrete context where the mathematical structure is present and could offer a framework in which to ask and answer relevant questions. The other consists of a purely *formal* understanding, where mathematical symbols and formulæ are memorized along with a set of "operational rules" to be used for manipulations. At this other extreme, the act of thinking about a problem is reduced to almost random attempts at manipulations in the hope of eventually obtaining "the correct answer" (which needs to be known or given for this approach to work). Even professional mathematicians can, at any given time, be operating in one of these two modes. Mathematics, however, is not pure intuition nor pure formalism, but the harmonious merging of the two, where form is supported by intuition and intuition is given form. Intuitions without muscles are mere conjectures just as thoughtless calculations are incoherent blabber. We all had teachers who were excellent at conveying the main ideas of a subject and others who put extreme emphasis on the rigor of presentations: I never was fully satisfied with any of these two approaches. I felt that they would lead to either ideas with no tools to implement them or tools with no idea on how to use them. So I always consciously attempted to design classes so as to incorporate enough of both. While I think I never really and fully succeeded, I gained innumerable insights that certainly improved my approach to teaching. This book is my latest attempt to reconcile the two souls of mathematics. The goal would remain hopelessly out of reach if I were to try to take a linear approach where all tools and concepts required to address

https://doi.org/10.1515/9783110780925-201

the problems, which will be discussed in the book, were first to be developed in a complete and rigorous way. It would indeed mean that I would have to write a book each about calculus, linear algebra, point set topology, analysis, probability, ordinary and partial differential equations, functional and harmonic analysis, geometry, etc. But I would rather like to showcase the power of mathematics by presenting the resolution of selected problems where it becomes apparent why, historically, the various basic mathematical subjects have developed. It is only when addressing concrete questions that one appreciates the remarkable unity of mathematics. Researchers will identify many unresolved issues inside any of the many mathematical branches and work tirelessly on resolving them, but mathematical subject areas were often born in response to the need of solving specific problems before they took on a life of their own. That is the reason why I take a problem and idea centered approach in this book. The hope is that the reader will never feel the need to ask: *why on earth are we doing this?* The problems I chose and the mathematics they engender in this presentation are not necessarily a historically correct account of the development of the corresponding ideas; they are more like a fictional history that could and might have been. They lead us on a plausible path from problem to mathematics as opposed to on a path from mathematics to application. I realize that this is a question of taste. My teaching experience tells me that this is what many students at least claim to crave for and what not too many textbooks offer. I also learned that mathematics is not necessarily easier to learn this way (and many students eventually settle for a more traditional presentation or learning mode once they realize it). I do not criticize the approach taken in the majority of available excellent textbooks. I would simply like to offer an alternative that can maybe prove useful to some students.

The target audience of this book is the advanced undergraduate or beginning graduate student who is interested in obtaining a transversal, if necessarily partial, view of the large edifice that is mathematics. As in most American universities real analysis is a subject that is often first learned as a graduate course; we avoided its use throughout the book. More often than not ideas cut across strict subject boundaries but are rarely presented in that way in standard courses. The latter are typically devoted to a single topic, which is systematically developed. In such an approach enlightening applications and/or connections to other areas can be absent or relegated to remarks. This books aims at offering an alternative and more inclusive perspective. The price that needs to be paid is that no topic can be as exhaustively covered as in a more traditional approach. The gain is the ability to present a selection of concepts and ideas that are central in mathematics and find use in a wide spectrum of areas ranging from pure to applied mathematics.

Prior exposure to calculus, linear algebra, and some probability is necessary for a full appreciation of the presentation. While the topics covered in this book intersect those covered in these courses, they are approached from a wider perspective that focuses on their fundamental motivations and they occasionally lead outside their traditional scope. The book is written in a colloquial style but maintains rigor every step of the way. It can be demanding for the reader since it encourages and requires the reader to think along and contribute to their own understanding. It may therefore occasionally be advisable to read it in a supervised mode.

The text is peppered with questions that the reader is invited to ponder to gain a fuller and deeper appreciation of the material. While the answers to many of such questions are a mere click away, the most effective reading requires full engagement. The reader is encouraged to work things out for themselves.

Enjoy the journey.

I would like to thank Herbert Amann, Sandro Merino, and Yucheng Zhang for their careful reading of early versions of the manuscript as well as for their constructive feedback that helped make this a better textbook.

Irvine, June 2022

1 Sets and functions

In this short chapter, we fix some basic but important notation in the hope of reducing later confusion or misunderstandings to a minimum. *Sets* will be viewed as collections of elements. The latter will be explicitly listed or described in some way. The set of natural numbers $\mathbb{N}$ is assumed to be given and understood

$$\mathbb{N} = \{1, 2, 3, \dots\}.$$

The prototypical set $\{1, 2, \dots, n\}$ of $n \in \mathbb{N}$ elements is denoted by $\mathbb{N}_n$. Other sets of numbers used in this text are $\mathbb{Q}$, $\mathbb{R}$, and $\mathbb{C}$ and will be discussed in the next chapter. If all elements of a set S also belong to another set T, we say that S is a *subset* of T or simply write $S \subset T$. Two sets S and T coincide if $S \subset T$ and $T \subset S$. Given two sets S and T, their *cross-product* $S \times T$ is defined as the set containing ordered pairs (s, t) with $s \in S$ and $t \in T$, namely

$$S \times T = \{(s, t) \mid s \in S, \ t \in T\}.$$

An example is the set $\mathbb{N} \times \mathbb{N}$ consisting of the ordered pairs (m, n) of integers m, n. The ordering condition ensures that $(1, 2)$ and $(2, 1)$ are distinct elements, i. e., that $(1, 2) \neq (2, 1)$. This stands in contrast to $\{1, 2\} = \{2, 1\}$.

Mathematics is made of definitions, statements, and proofs. They try to capture observations, intuitions, and structures with concepts and their provable properties and connections. Statements can sometimes be related to each other by implication, meaning that validity of one follows from the validity of another. The notation $A \Rightarrow B$ is used to indicate that the validity of A entails that of B. When two statements are equivalent, we write $A \Leftrightarrow B$ (read as "A if and only if B" and sometimes also written as "A iff B"). Equivalence of the statements A and B is the same as the combination of $A \Rightarrow B$ and $B \Rightarrow A$. The negation of a statement A is denoted by $\neg A$. When attempting to prove $A \Rightarrow B$, we can just as well try to prove the equivalent $\neg B \Rightarrow \neg A$.

A *map* $m : D \subset S \to T$ between two sets S and T is a triple (D, T, M) consisting of a *domain of definition* (domain) $D \subset S$, of a target set T, and of a subset M of their cross-product $S \times T$ satisfying:

($\mathbf{m_1}$) $D = \{s \in S \mid (s, t) \in M \text{ for some } t \in T\}$

($\mathbf{m_2}$) For each $s \in D$, there is a unique $t \in T$ such that $(s, t) \in M$.

If condition ($\mathbf{m_2}$) holds, we simply write $t = m(s)$, meaning that t is the *value* of the map m at s. The set M is called the *graph* of the map and is sometimes denoted by $G(m)$. The elements of the domain of definition D of a map m are called the *arguments of the map*. The collection of elements t of T, which occur as values of a map m, that is, such that an $s \in D$ can be found with $m(s) = t$ is called the *range* of the map m and is denoted by R. To avoid confusion, we will sometimes include the map m in the

https://doi.org/10.1515/9783110780925-001

notation of its domain and range and write $D(m)$ and $R(m)$. If a map $m : D \subset S \to T$ is such that $D = S$, we simply write $m : S \to T$.

As a simple example of a map, consider a store S as the set of all items it contains. Not all items found in the store will be for sale. Cash registers are examples of such items. We denote the collection of all items for sale by D and associate to them their sale price. In this way, we obtain a map

$$p : D \subset S \to [0, \infty), \quad \text{item} \mapsto p(\text{item}),$$

the item pricing map. This example shows the intuitive nature of the defining properties of a map. It is indeed desirable to limit sales to the items actually available for purchase and to have a unique price for each such item. Given a map $m : S \to T$ between two sets S, T and a subset $V \subset T$, the *preimage* $m^{-1}(V)$ of V under m is defined as the set

$$m^{-1}(V) = \{s \in S \mid m(s) \in V\} \subset S.$$

Given $U \subset S$, it is sometimes useful to consider the *image set* $m(U)$ of U under m that is given by

$$m(U) = \{m(s) \mid s \in U\} \subset T.$$

Notice that $R(m) = m(S)$. The *restriction* $m|_U$ of m to the subset U is defined as the (different) map

$$m|_U : U \to T, \quad s \mapsto m(s).$$

The whole field of mathematics can be viewed as an effort to understand specific maps. Maps can be distinguished by their properties, the most basic of which are *injectivity*, *surjectivity*, and *bijectivity*. A map $m : D \to T$ is called injective, or *one-to-one* if it holds that

$$m(r) = m(s) \quad \text{for } r, s \in D \Longrightarrow r = s.$$

In plain English, a map is one-to-one if each and any of its value $t \in R$ is taken on at most at a single argument $s \in D$, or, equivalently, iff different arguments always yield different values, that is, iff

$$r, s \in D, \quad r \neq s \Longrightarrow m(r) \neq m(s)$$

A map is called surjective, or *onto*, if it holds that $R(m) = T$. This means that each element of the target space occurs as a value of the map, i. e., that

$$\text{for each } t \in T \text{ there is } s \in S \text{ s. t. } m(s) = t,$$

where s. t. stand for *such that*. Finally, a map $m : D \subset S \to T$ is bijective, iff it is both injective and surjective, or, equivalently iff

$$\text{for each } t \in T \text{ there is a unique } s \in D \text{ s. t. } m(s) = t.$$

If two maps $m : S \to T$ and $n : U \to V$ between sets S, T, U, V are such that $m(S) \subset U$, then they can be composed to obtain their *composition* $n \circ m$ given by

$$n \circ m : S \to V, \quad s \mapsto n(m(s)).$$

If a given map $m : S \to T$ is bijective, it admits an inverse, denoted by $m^{-1} : T \to S$[1] such that

$$m^{-1}(m(s)) = s \text{ for } s \in S \quad \text{and} \quad m(m^{-1}(t)) = t \text{ for } t \in T.$$

This means that

$$m^{-1} \circ m = \mathrm{id}_S \quad \text{and} \quad m \circ m^{-1} = \mathrm{id}_T$$

where id_U denotes the identity map of the set U,

$$\mathrm{id}_U : U \to U, \quad s \mapsto s.$$

What is the graph of the inverse of a function in terms of the graph of the function itself? (Q1)

As a simple example, consider the following map:

$$f = (\mathbb{R}, \mathbb{R}, \{(x, x^2) \mid x \in \mathbb{R}\} = G(f)),$$

which we will more simply denote as

$$f : \mathbb{R} \to \mathbb{R}, \quad x \mapsto x^2,$$

where $m = f$, $D = S = \mathbb{R}$, and $T = \mathbb{R}$. We intentionally avoid the notation $f(x) = x^2$ because it quickly can lead to confusion (as it does not give information about the domain of definition nor about the target set). To drive this point home, we consider also the following other two functions:

$$g : \mathbb{R} \to [0, \infty), \quad x \mapsto x^2,$$
$$h : [0, \infty) \to [0, \infty), \quad x \mapsto x^2,$$

1 Notice that in the earlier definition of preimage we used the same notation even though the map may not have been invertible. The distinction is that preimages are taken of sets and not of individual elements.

which are obtained from f by modifying its target set and both domain and target sets, respectively. These could loosely be referred to as the function x^2. It, however, holds that f is not one-to-one nor onto, while g is onto but not one-to-one, and h is actually one-to-one and onto, and has an inverse h^{-1} given by

$$h^{-1} : [0, \infty) \to [0, \infty), \quad x \mapsto \sqrt{x}.$$

You can see how easily we could run into trouble by giving up mathematical precision in the name of simplicity of notation and expediency. Mathematicians are not always this "pedantic" about details but this is due to the fact that they are fully aware of the details they swipe under the rug. If you are still learning mathematics, we advise that you accept this pedantry as the price to pay for a clean understanding of the concepts and in order to avoid confusion later when dealing with more sophisticated concepts and more involved situations.

A set S is called *finite* if there is an integer $n \in \mathbb{N}$ and a bijective map $c : \mathbb{N}_n \to S$. You can think of c as a way of counting the elements of S, where $c(1)$ is the first and $c(n)$ the last. Any set, which turns out not to be finite, is called *infinite*. Among infinite sets, those which can be represented as the range of a bijective map defined on $\mathbb{N}$ are called *countable (countably infinite)*. Examples are the set of integers $\mathbb{Z}$, that of rational numbers $\mathbb{Q}$, or the subset of even numbers $2\mathbb{N} = \{2n \mid n \in \mathbb{N}\} \subset \mathbb{N}$.

It is sometimes possible to introduce operations on a set. By an operation $*$ on a set S, we mean a map

$$* : S \times S \to S, \quad (r, s) \mapsto *(r, s) = r * s,$$

which combines two elements of the set to yield a third. In this case, we refer to S with the operation $*$ as $(S, *)$. The operation $*$ is called *commutative* if it holds that

$$*(r, s) = r * s = s * r = *(s, r) \quad \text{for each } r, s \in S.$$

It is called *associative* if it holds that

$$(r * s) * t = r * (s * t) \quad \text{for each } r, s, t \in S.$$

It is said to have an *identity element* if there is $e \in S$ with

$$e * s = s * e = s \quad \text{for each } s \in S.$$

If $(S, *)$ admits an identity element e, an element $s \in S$ is said to have an *inverse* (element) if $t \in S$ can be found satisfying

$$t * s = s * t = e.$$

Such an inverse element is sometimes called s^{-1}. $(S,*)$ is called a *group* if its operation $*$ is associative and admits an identity element as well as an inverse for each of its elements. If commutativity holds, then we speak of a *commutative group*. A set $(S,+,\cdot)$ with two operations $+$ and $\cdot$ such that $(S,+)$ and $(S^* = S\setminus\{e\},\cdot)$, where e is the identity element of $(S,+)$, are commutative groups, is called a *field* if it additionally satisfies *distributivity*, i. e., if

$$r\cdot(s+t) = r\cdot s + r\cdot t \quad \text{for each } r,s,t\in S.$$

In a field, $+$ is called addition, $\cdot$ multiplication, the additive identity is denoted by 0, and the multiplicative identity by 1.

Letting $\mathbb{F}=\mathbb{R},\mathbb{C}$ or your favorite field, we consider maps $v:\mathbb{N}_n\to\mathbb{F}$. Since the domain of definition consists of the first n integers, any such map is fully determined by its values $v(1),\dots,v(n)\in\mathbb{F}$, which are from now on referred to as $v_1,\dots,v_n$. The map v can therefore be viewed as the n-vector (n-tuple),

$$(v_1,\dots,v_n)\in\mathbb{F}^n = \underbrace{\mathbb{F}\times\cdots\times\mathbb{F}}_{n\text{ times}}.$$

One can also consider maps $x:\mathbb{N}\to\mathbb{F}$, which are similarly fully determined by their values $x(n)$ for $n\in\mathbb{N}$. In this case, the values are simply denoted by x_n, $n\in\mathbb{N}$, and the map is thought of as the sequence

$$(x_n)_{n\in\mathbb{N}} = (x_1,x_2,x_3,\dots)$$

in the field $\mathbb{F}$. The field $\mathbb{F}$ can clearly be replaced by any set S and the domain $\mathbb{N}$ by $\mathbb{N}_n$ to obtain sequences and n-tuples of elements of the set S, respectively. Just like it is often convenient to think of vectors as functions, it is sometimes useful to think of functions as vectors, "very long vectors" at that. Let S be any set and consider functions $f:S\to\mathbb{F}$. We can think of the values $f(s)$ as the components of the vector $(f(s))_{s\in S}$. Notice that, since the values are taken in a field $\mathbb{F}$, they can be added and multiplied. This makes it possible to add and multiply functions as well through

$$\begin{aligned}(f+g)(s) &= f(s)+g(s) \quad \text{for } s\in S,\\ (f\cdot g)(s) &= f(s)\cdot g(s) \quad \text{for } s\in S,\end{aligned}$$

for $f,g\in\mathbb{F}^S$, where $\mathbb{F}^S$ denotes the collection of all functions $f:S\to\mathbb{F}$. With this notation, we recognize that $\mathbb{F}^n = \mathbb{F}^{\mathbb{N}_n}$. We can also, for instance, think of complex sequences $(z_n)_{n\in\mathbb{N}}$ as vectors in $\mathbb{C}^{\mathbb{N}}$ or of real real-valued functions $f:\mathbb{R}\to\mathbb{R}$ as vectors in $\mathbb{R}^{\mathbb{R}}$. In general, we may sometimes denote the set or collection of all maps defined on a set S with values in a set T by T^S. The notation is motivated by the fact that, if S and T are finite sets and $|R|$ denotes the number of elements of a finite set R, then it holds that

$$|T^S| = |T|^{|S|}.$$

Given a set S, finite or infinite, a collection of subsets $\mathcal{S} = \{S_i \mid i \in I\}$, where I is an index set (not necessarily finite), is called a *partition* of the set S if it holds that:

(p1) $S_i \neq \emptyset$ for $i \in I$.

(p2) $S_i \cap S_j = \emptyset$, whenever $i \neq j$ for $i, j \in I$.

(p3) $\bigcup_{i \in I} S_i = S$.

Here, $\emptyset$ denotes the empty set. A partition therefore consists of subsets, which each contribute at least one element of S, which do not contribute an element of S more than once, and which exhaust all elements of S.

For the purpose of simplifying notation, we shall sometimes use quantifiers in statements. When writing $\forall$, we mean *for every*, so that $\forall\, n \in \mathbb{N}$ means for each and every natural number. A complete example would be "$2n$ is even $\forall\, n \in \mathbb{N}$" to mean "$2n$ is an even number for each natural number $n \in \mathbb{N}$". The expression $\exists$ means *there exists*, whereas $\exists!$ means *there exists a unique*, so that "$\forall \varepsilon \in \mathbb{R}$ with $\varepsilon > 0\ \exists\, n \in \mathbb{N}$ s. t. $\frac{1}{n} < \varepsilon$" means "that each positive real number admits a rational number of the form $\frac{1}{n}$ that is smaller than it".

2 Numbers

While we assume that natural numbers are given, we will construct whole, rational, real, and complex numbers from them in this chapter. Doing so will allow us to introduce concepts that are useful in many other subject areas within mathematics. Along with the natural numbers $\mathbb{N}$, we assume that the two operations of addition and multiplication

$$+ : \mathbb{N} \times \mathbb{N} \to \mathbb{N}, \quad (m, n) \mapsto m + n,$$
$$\cdot : \mathbb{N} \times \mathbb{N} \to \mathbb{N}, \quad (m, n) \mapsto m \cdot n,$$

are given and satisfy the properties we all learned in grade school as arithmetic (commutativity and associativity for both, the existence of an identity for multiplication and distributivity).

2.1 Rational numbers

It is likely a historical fiction but still conceivable that numbers have been "invented" to satisfy our need to measure and quantify. Since there are no absolute quantities, the first and rougher estimates of quantity are *more* and *less*. If we care to be more precise, however, we could and do understand size in terms of arbitrarily chosen *units*. The first definition of *one meter (m)* was, for instance, the length of a specific bar, which was held in Paris. Measurement then relates the object to be measured back to the chosen reference unit. With the natural numbers at our disposal, we can understand objects whose length is an integer multiple of the unit. But what about objects, which cannot be measured exactly as an integer number of units such as your height in meters (most likely)? This question points to the need of finding a way to express sizes that go beyond simple integer multiples of a unit. Using the example of a meter, we can come across an object that fits exactly two times in a meter. In more abstract terms, this relationship could be captured as *it takes two (of these) to make one unit*, which again only requires the use of two natural numbers to express and can be taken as the definition of the word one-half or of the number $\frac{1}{2}$. How can we turn this simple idea into a proper mathematical definition of rational numbers? Consider the problem of obtaining a construction of the number $\frac{m}{n}$ for $m, n \in \mathbb{N}$. It would correspond to the "real world" relationship *it takes n (of these) to make m units*. Again, describing what will eventually be $\frac{m}{n}$, only requires access to $m, n \in \mathbb{N}$. We could therefore simply stipulate that $\frac{m}{n}$ is the pair (m, n) and the new set of these numbers to be

$$\mathbb{N} \times \mathbb{N} = \{(m, n) \mid m, n \in \mathbb{N}\}.$$

If we did that, however, *it takes n to make m units* and *it takes 2n to make 2m units* would be different descriptions of the same underlying relationship, and hence arguably of

https://doi.org/10.1515/9783110780925-002

the same number we are trying to define. In other words, we would have to conclude that $(m, n) = (2m, 2n)$, but these pairs are ostensibly not identical.

At this point, we introduce and use the first "mathematical trick" (device, idea), which turns out to be ubiquitous in mathematics: we form so-called equivalence classes. Given numbers $m, n \in \mathbb{N}$, we set

$$[m, n] = \{(\overline{m}, \overline{n}) \mid m \cdot \overline{n} = \overline{m} \cdot n\},$$

so that, e. g., $(m, n) \in [m, n]$ but also $(2m, 2n) \in [m, n]$, since $2m \cdot n = m \cdot 2n$ thanks to the properties of multiplication of the naturals. Any specific pair $(m, n) \in \mathbb{N}$ determines, and hence belongs to one and only one of these sets. These do in fact build a partition of the set $\mathbb{N} \times \mathbb{N} = \mathbb{N}^2$ into disjoint and nonempty subsets. Any element of one of these sets encodes the same "factual" relationship *it takes this many to make that many units*. This idea of condensing certain subsets into a single element is based on the mathematical construct of *equivalence relation*. Given a set S, an equivalence relation on it is a map $\sim: S \times S \to \{0, 1\}$ satisfying:

(er1) $\sim (s, s) = 1$ for $s \in S$.
(er2) If $\sim (r, s) = 1$, then $\sim (s, r) = 1$ for $r, s \in S$.
(er3) If, for $r, s, t \in S$, it holds that $\sim (r, s) = 1$ and $\sim (s, t) = 1$, then $\sim (r, t) = 1$.

It is customary to write $r \sim s$ and $r \nsim s$ instead of the lengthier $\sim (r, s) = 1$ and $\sim (r, s) = 0$, respectively. These expressions are read as *r is in relation with s* and *not in relation with s*, respectively. In the specific setting above, we would define an equivalence relation on $\mathbb{N}^2$ by setting

$$(m, n) \sim (\overline{m}, \overline{n}) \quad \text{iff} \quad m \cdot \overline{n} = \overline{m} \cdot n,$$

for any given $(m, n), (\overline{m}, \overline{n}) \in \mathbb{N}^2$.

Next, given a set S and an equivalence relation $\sim$ on it, we define

$$[\![s] = \{r \in S \mid r \sim s\} \quad \text{and} \quad [s]\!] = \{r \in S \mid s \sim r\},$$

the sets of elements left- and right-related to s, respectively. In view of condition **(er1)**, it holds that $s \in [\![s]$ and $s \in [s]\!]$, and thanks to **(er2)** also that $[\![s] = [s]\!]$. We will therefore simply write $[s]$ instead of $[\![s]$ or $[s]\!]$. The last condition **(er3)** ensures that

$$s \in [r],\ r \in [t] \Longrightarrow s \in [t]$$

Finally, notice that $[r] \cap [t] \neq \emptyset$ implies that $[r] = [t]$. Convince yourself that the set of all equivalence classes

$$S/{\sim} = \{[s] \mid s \in S\}$$

yields a partition (Q1) of the set S and that the relation defined above on $\mathbb{N}^2$ is indeed an equivalence relation (Q2), i. e., that it satisfies conditions **(er1)**–**(er3)**. After these preparations, we can define

$$\mathbb{Q}^+ = \mathbb{N}^2/\sim,$$

the set of positive rational numbers.

At this point, we introduce another universally important concept. While one is free to make any definition, one is also required to check that the concept introduced is *well-defined* meaning that it is unambiguous and noncontradictory, that it does not alter the meaning of any previous concept it extends, and that it admits at least one example (existence of the described objects). In the case of the definition of $\mathbb{Q}^+$, this means first and foremost that any two subsets of $\mathbb{N}^2$ determined by $\sim$, supposedly defining a rational number, do either coincide or have empty intersection. This can be proven to be the case by showing that $\sim$ is indeed an equivalence relation (the very properties of equivalence relation are introduced for that purpose). If that were not the case, there would be different rational numbers, which share the same representative, thus leading to some confusion indeed. We also need to verify that the new numbers are compatible with the natural numbers, in the sense that the latter can be viewed as a subset of the former, i. e., that

$$\mathbb{N} \subset \mathbb{Q}^+ = \mathbb{N}^2/\sim.$$

To do so, consider the special equivalence classes $[m, 1]$ for $m \in \mathbb{N}$, which encode the relation *it takes 1 of these to make m units* and clearly coincides with m units. This is summarized by the map

$$\iota : \mathbb{N} \to \mathbb{Q}^+ = \mathbb{N}^2/\sim, \quad m \mapsto [m, 1]$$

being injective (one-to-one). Natural numbers come with two natural structures: they can be added together and multiplied with one another. It is therefore natural to try and extend these operations to $\mathbb{Q}^+$. When doing so, we first need to define addition and multiplication on $\mathbb{Q}^+ \times \mathbb{Q}^+$. We take two elements $p = [m, n]$ and $q = [k, l]$, which are two sets (equivalence classes), and look for ways to define $p+q$ and $p \cdot q$. Given that an equivalence class is determined by any of its members, we choose representatives $(\widetilde{m}, \widetilde{n})$ and $(\widetilde{k}, \widetilde{l})$ for p and q, respectively. Considering the product first, we may define

$$p \cdot q = [\widetilde{m} \cdot \widetilde{k}, \widetilde{n} \cdot \widetilde{l}].$$

This is motivated by considering an example. Notice that $n \cdot [m, n] = m$, since $[m, n]$ means *it takes n to make m units*. This can be rewritten as

$$[n, 1] \cdot [m, n] = [m, 1].$$

Now observe that $[m, 1] = [n \cdot m, 1 \cdot n]$, which is compatible with the proposed definition for the product. It still remains to verify, however, that this product is well-defined. In this case, this amounts to showing that, while we use representatives in the defining expression, the resulting equivalence class for the product does not depend on the choice of these representatives. To that end, we pick $(\overline{m}, \overline{n}) \sim (\widetilde{m}, \widetilde{n})$ and $(\overline{k}, \overline{l}) \sim (\widetilde{k}, \widetilde{l})$. This means that $\overline{m} \cdot \widetilde{n} = \overline{n} \cdot \widetilde{m}$ and that $\overline{k} \cdot \widetilde{l} = \overline{l} \cdot \widetilde{k}$. It follows that

$$(\overline{m} \cdot \widetilde{n}) \cdot (\widetilde{k} \cdot \overline{l}) = (\widetilde{m} \cdot \overline{n}) \cdot (\overline{k} \cdot \widetilde{l}),$$

which rewrites as

$$(\overline{m} \cdot \overline{k}) \cdot (\widetilde{n} \cdot \widetilde{l}) = (\widetilde{m} \cdot \widetilde{k}) \cdot (\overline{n} \cdot \overline{l}).$$

This, in turn means that $(\overline{m} \cdot \overline{k}, \overline{n} \cdot \overline{l}) \sim (\widetilde{m} \cdot \widetilde{k}, \widetilde{n} \cdot \widetilde{l})$ or simply that

$$[\widetilde{m} \cdot \widetilde{k}, \widetilde{n} \cdot \widetilde{l}] = [\overline{m} \cdot \overline{k}, \overline{n} \cdot \overline{l}].$$

We conclude that the definition of product does not depend on the necessary choice of representatives. Notice that we used the standard properties of integer multiplication, such as commutativity and associativity, in this argument. It can be shown (consider this an exercise) that this multiplication of (positive) rationals not only preserves all the properties of integer multiplication (commutativity, associativity, and the existence of an identity element), but gains some new. In particular, each $p = [m, n]$ will have a multiplicative inverse, i. e., a rational number q satisfying $p \cdot q = q \cdot p = 1 = [1, 1]$. What is a representative of its equivalence class? (Q3)

Next, we would like to investigate the possibility of adding positive rational numbers. How can we define $[m, n] + [k, l]$ in a way that is compatible with the addition of natural numbers and unambiguous, i. e., well-defined? We again start with the intuition behind the definition of rational numbers. The rationals $[m, n]$ and $[k, l]$ encode *it takes n to make m units* and *it takes l to make k units*, respectively. Thus if we take $n \cdot l$ many of $[m, n]$ and of $[k, l]$, we must end up with $l \cdot m + n \cdot k$ units. Thus it "should" hold that

$$n \cdot l \cdot ([m, n] + [k, l]) = l \cdot m + n \cdot k.$$

This motivates the definition of addition by

$$[m, n] + [k, l] = [\widetilde{l} \cdot \widetilde{m} + \widetilde{n} \cdot \widetilde{k}, \widetilde{n} \cdot \widetilde{l}]$$

for $(\widetilde{m}, \widetilde{n}) \in [m, n]$, $(\widetilde{k}, \widetilde{l}) \in [k, l] \in \mathbb{Q}^+$. It remains to verify that this definition is self-consistent, in the sense that it does not depend on the choice of representatives used. Take $(\overline{m}, \overline{n}) \in [m, n]$ and $(\overline{k}, \overline{l}) \in [k, l]$: we need to show that

$$[\widetilde{l} \cdot \widetilde{m} + \widetilde{n} \cdot \widetilde{k}, \widetilde{n} \cdot \widetilde{l}] \sim [\overline{l} \cdot \overline{m} + \overline{n} \cdot \overline{k}, \overline{n} \cdot \overline{l}]. \tag{2.1}$$

By definition of the equivalence relation, we have that

$$(\overline{m},\overline{n}) \sim (\widetilde{m},\widetilde{n}) \text{ and } (\overline{k},\overline{l}) \sim (\widetilde{k},\widetilde{l}) \Longleftrightarrow \overline{m}\cdot\widetilde{n} = \overline{n}\cdot\widetilde{m} \text{ and } \overline{k}\cdot\widetilde{l} = \overline{l}\cdot\widetilde{k},$$

and, consequently, that

$$\begin{aligned}(\overline{l}\cdot\overline{m}+\overline{n}\cdot\overline{k})\cdot(\widetilde{n}\cdot\widetilde{l}) &= \overline{l}\cdot\widetilde{l}\cdot\overline{m}\cdot\widetilde{n}+\overline{n}\cdot\widetilde{n}\cdot\overline{k}\cdot\widetilde{l}\\ &= \overline{l}\cdot\widetilde{l}\cdot\overline{n}\cdot\widetilde{m}+\overline{n}\cdot\widetilde{n}\cdot\overline{l}\cdot\widetilde{k}\\ &= (\widetilde{l}\cdot\widetilde{m}+\widetilde{n}\cdot\widetilde{k})\cdot(\overline{n}\cdot\overline{l}),\end{aligned}$$

and thus, again by definition of the equivalence relation, that indeed (2.1) holds. Next, we verify that addition of rationals is an extension of addition for the naturals. Since $m = [m,1]$ and $n = [n,1]$ by the identification of $\mathbb{N}$ with a subset of $\mathbb{Q}^+$, we see that

$$[m,1] + [n,1] = [m\cdot 1 + n\cdot 1, 1\cdot 1] = [m+n,1] = m+n,$$

for any $m, n \in \mathbb{N}$.

Once positive rational numbers are understood, we can start using a simpler but more suggestive notation

$$[m,n] = \frac{m}{n},$$

which could generate potential ambiguities but is much more convenient in algebraic manipulations. Similarly, we stipulate that

$$\mathbb{Q}^+ = \left\{\frac{m}{n}\;\middle|\; m,n\in\mathbb{N}\right\}.$$

2.2 Whole numbers

The mathematical construction of whole numbers starting from the naturals is fully analogous to that of positive rational numbers. Still, intuition and motivation may help understanding. While assumed to be a given set, no questions asked, we tend to think of naturals as a sequence. More precisely, we have an *order* on them, denoted by $\leqslant$, which makes it possible to compare any two natural numbers n and m and determine whether one is larger than the other. If $m \leqslant n$ and $n \leqslant m$, we have that $m = n$ and will therefore write $m < n$ if it so happens that $m \leqslant n$ and $m \neq n$.

Notice that we are not giving a formal definition of order here, but still encourage the reader to think about finding just such a definition in order to practice your translational abilities between intuition and mathematical language.

Given $m, n \in \mathbb{N}$ with $m < n$, it is possible to solve the equation $m + x = n$ for $x \in \mathbb{N}$, i. e., to find a natural number $k \in \mathbb{N}$ such that $m + k = n$. Thus we can think of m as being k units *ahead* of n in the sequence of naturals. If, on the other hand, $m > n$, the same equation $m + x = n$ cannot be solved in the realm of natural numbers, but there still is $k \in \mathbb{N}$ such $n + k = m$; in other words, it is now n that is k units ahead of m. We could encode the relationship between any two natural numbers m and n by simply recording both numbers in the pair (m, n). This expression would mean that m is k units ahead of n, if $m + k = n$, and m is k units past n, if $n + k = m$. At this point, we would notice that $(m + l, n + l)$ encodes that same relationship as (m, n), thus leading to ambiguity in the definition of the relationship. The concept of equivalence relation again comes to the rescue. We will say that

$$(m, n) \sim (k, l) \iff m + l = n + k,$$

whenever $m, n, k, l \in \mathbb{N}$ are given. Notice that this definition only requires to have and understand natural numbers and uses addition of naturals, that we also assume to be given. The reader is encouraged to verify that this is indeed an equivalence relation (Q4) on $\mathbb{N}^2$. It is then possible to introduce a new set of numbers, which extends the naturals, by setting

$$\mathbb{Z} = (\mathbb{N}^2/\sim) \ni [m, n],$$

consisting of the equivalence classes $[m, n]$ determined by the above equivalence relation. The set $\mathbb{N}$ can be viewed as a subset of $\mathbb{Z}$ via the injective map

$$\iota : \mathbb{N} \to \mathbb{N}^2, \quad n \mapsto [n + 1, 1],$$

so that, e. g., 1 is identified with $[2, 1]$, thus corresponding to the unit that separates 2 from 1. The mapping ι is not surjective and new numbers are created that do not belong to $\mathbb{N}$. Among them is $[1, 1]$.

Addition on this new set can be defined by

$$[m, n] + [k, l] = [m + k, n + l], \quad m, n, k, l \in \mathbb{N},$$

and it still enjoys all the properties of addition for naturals. We leave it as an exercise (Q5) to verify that this definition makes sense (read *is well-defined*), that it extends addition of naturals, and that commutativity and associativity are preserved.

The special new number $[1, 1]$ enjoys the property that

$$[m, n] + [1, 1] = [m + 1, n + 1] = [1, 1] + [m, n] = [m, n],$$

for any $m, n \in \mathbb{N}$, i. e., for any $[m, n] \in \mathbb{Z}$. This number usually goes by the name of 0 and is the identity element of addition in the new set $\mathbb{Z}$. Each natural number $n = [n + 1, 1]$ has a “companion” number in $\mathbb{Z}$ given by $[1, n + 1]$. Notice that

$$[n + 1, 1] + [1, n + 1] = [n + 2, n + 2] = [1, 1] = 0,$$

so that $[1, n+1]$ is the additive inverse (the negative) of n and is from now on denoted by $-n$. The equivalence classes determining the whole numbers can be visualized as the diagonals in the depiction below, where the corresponding whole number is indicated in blue.

	1	2	3	4	5	6	7	8
0	(0, 0)	(1, 0)	(2, 0)	(3, 0)	(4, 0)	(5, 0)	(6, 0)	(7, 0)	(8, 0)
-1	(0, 1)	(1, 1)	(2, 1)	(3, 1)	(4, 1)	(5, 1)	(6, 1)	(7, 1)	(8, 1)
-2	(0, 2)	(1, 2)	(2, 2)	(3, 2)	(4, 2)	(5, 2)	(6, 2)	(7, 2)	(8, 2)
-3	(0, 3)	(1, 3)	(2, 3)	(3, 3)	(4, 3)	(5, 3)	(6, 3)	(7, 3)	(8, 3)
-4	(0, 4)	(1, 4)	(2, 4)	(3, 4)	(4, 4)	(5, 4)	(6, 4)	(7, 4)	(8, 4)
-5	(0, 5)	(1, 5)	(2, 5)	(3, 5)	(4, 5)	(5, 5)	(6, 5)	(7, 5)	(8, 5)
-6	(0, 6)	(1, 6)	(2, 6)	(3, 6)	(4, 6)	(5, 6)	(6, 6)	(7, 6)	(8, 6)
-7	(0, 7)	(1, 7)	(2, 7)	(3, 7)	(4, 7)	(5, 7)	(6, 7)	(7, 7)	(8, 7)
-8	(0, 8)	(1, 8)	(2, 8)	(3, 8)	(4, 8)	(5, 8)	(6, 8)	(7, 8)	(8, 8)

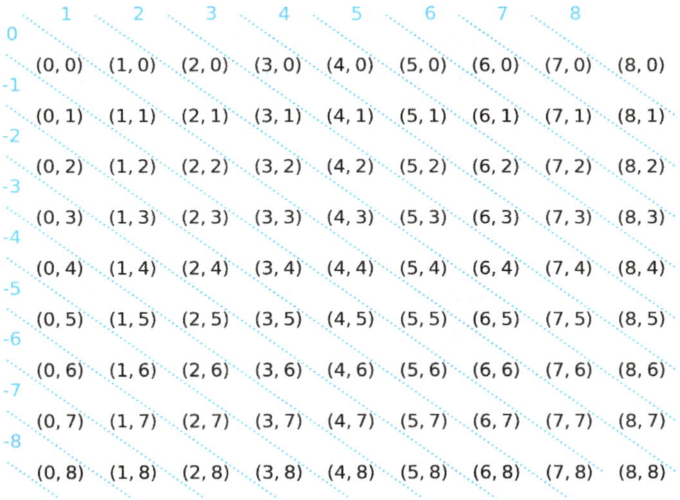

It holds that

$$n + (-n) = 0 = (-n) + n, \quad n \in \mathbb{N},$$

and $n + (-m) = (-m) + n$ is simply denoted by $n - m = -m + n$ for $m, n \in \mathbb{N}$. Addition and multiplication on $\mathbb{N}$ also satisfy the property of distributivity

$$m \cdot (k + l) = m \cdot k + m \cdot l, \quad m, k, l \in \mathbb{N}.$$

Verify that this property is also valid in $\mathbb{Z}$ and $\mathbb{Q}^+$. Notice that you need to define multiplication (Q6) of whole numbers, in particular. We conclude this section by suggesting that the reader give an explicit construction (Q7) of the full set of rationals $\mathbb{Q}$ either starting with $\mathbb{Z}$ or with $\mathbb{Q}^+$. In the process, show that the starting set of numbers can be identified with a subset of the new set and that addition, multiplication, as well as order structure can all be extended to the new set. From now on, we will no longer use $\cdot$ to denote multiplication of numbers.

2.3 Real numbers

2.3.1 Construction of the reals I

In the previous sections, we convinced ourselves that whole and rational numbers can be thought of as equivalence classes of pairs of integers. While this aspect was not stressed, it can also be argued that whole numbers are constructed in an attempt to make sure that equations of the type

$$m + x = n$$

always possess a unique solution x (does not need to be possible but turns out to be), while rational numbers are introduced in an effort to solve multiplicative equations of type $m \cdot x = n$. This is not only possible but it can be done in finitely many steps in the sense that it is sufficient to consider pairs of naturals or pairs of whole numbers to craft an intuitive and formal understanding of whole and rational numbers, respectively. This is, however, not possible when it comes to the reals. The search for the latter can be motivated by the quest for solutions of algebraic equations, of which $x^2 = 2$ is prototypical. The first step is to show that $\mathbb{Q}$ is insufficient to allow for solutions of this equation and the second will consist in giving a natural argument for the necessity of additional numbers.

The classical argument that shows $q^2 \neq 2$ for $q \in \mathbb{Q}$ is by contradiction.[1] Assume that $q = \frac{m}{n}$, that m and n do not share any factors, and that $q^2 = 2$. Then $m^2 = 2n^2$, which makes m^2 an even number (defined as a multiple of 2). For m^2 to be even, though, m needs to be as well: indeed if 2 is not a prime factor of m, it cannot possibly be one of m^2 (here we tacitly use the unique factorization theorem for integer numbers in a product of integer powers of primes). Then $m = 2k$ for some $k \in \mathbb{N}$, which implies

$$4k^2 = m^2 = 2n^2.$$

It follows that $n^2 = 2k^2$, which similarly makes n into an even number and violates the assumption that m and n do not share any prime factors. We conclude that $q^2 \neq 2$ for all $q \in \mathbb{Q}$.

Next, consider the function $f : \mathbb{Q} \to \mathbb{Q}, x \mapsto x^2 - 2$, the graph of which is depicted below.

1 A proof is said to be by contradiction if it starts by assuming the negation of the statement to be proved and concludes by deriving a contradiction from it, and thus showing that the statement must actually be true.

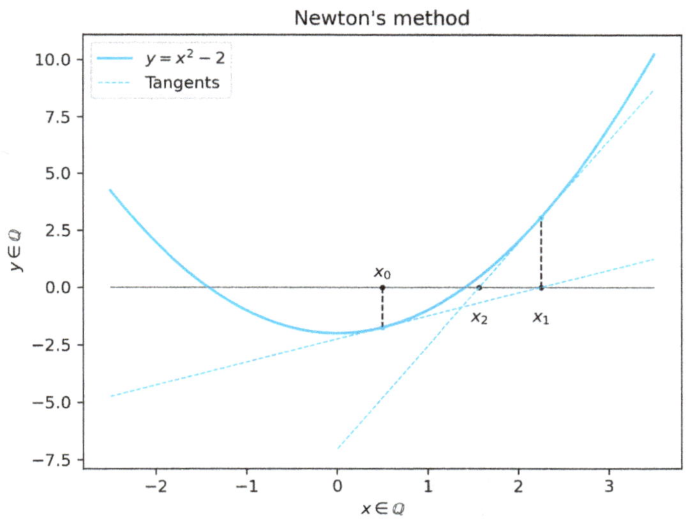

If we are interested in determining what appears to be a possible positive zero of f (we know it does not exists in $\mathbb{Q}$), a strategy we could follow would be to start with an initial guess $x_0 > 0$ as depicted, then replace the actual function f with its tangent at the point $(x_0, f(x_0))$, determine its zero x_1, and continue in this fashion to obtain $x_2, x_3, \ldots$. The plot seems to indicate that x_j moves closer to the point of intersection of the graph of f with the x-axis. In order to turn this idea into an algorithm, we need to determine the tangent lines to the graph of f wherever needed. Fix $0 < x_0 \in \mathbb{Q}$. A line l_{x_0} through the point $(x_0, f(x_0) = x_0^2 - 2)$ has necessarily the form

$$l_{x_0}(x) = m(x - x_0) + f(x_0) = m(x - x_0) + x_0^2 - 2.$$

We therefore try to determine m so that f and l_{x_0} have a single point in common, i. e., such that the quadratic equation

$$0 = x^2 - 2 - m(x - x_0) - x_0^2 + 2 = \left(x - \frac{m}{2}\right)^2 - \frac{m^2}{4} + mx_0 - x_0^2$$

has a single solution. This is the case only if

$$m^2 - 4mx_0 + 4x_0^2 = (m - 2x_0)^2 = 0,$$

that is, if $m = 2x_0$. It follows that x_1 is determined by the equation $2x_0(x-x_0)+x_0^2-2 = 0$, which gives

$$x_1 = x_0 + \frac{2 - x_0^2}{2x_0} = \frac{x_0}{2} + \frac{1}{x_0}.$$

As the initial point is positive but otherwise arbitrary and $x_1 > 0$ if $x_0 > 0$, we see that $x_2 = \frac{x_1}{2} + \frac{1}{x_1}$. We therefore obtain the recursively defined sequence $(x_n)_{n\in\mathbb{N}} \in \mathbb{Q}^{\mathbb{N}}$,

$$\begin{cases} x_0 > 0 \\ x_{n+1} = \frac{x_n}{2} + \frac{1}{x_n}, \quad n \geqslant 1. \end{cases} \tag{2.2}$$

Before we continue the discussion toward the introduction of real numbers, we take a detour to highlight the general nature of the procedure we followed to obtain the above sequence.

2.3.2 Newton's method

Given a function $f : V \to W$ between two vector spaces (possibly infinite dimensional and defined later in Chapter 3) and the equation $f(x) = 0$, one can apply the same idea as above: pick an initial guess x_0 and replace the function f by its linearization ("tangent") to produce a hopefully better approximation of an actual zero of f. In order to do this, one needs the function f to satisfy some differentiability condition (see Chapter 5). Denoting the derivative of f at x_0 by $Df(x_0)$, this amounts to solving

$$f(x) \simeq f(x_0) + Df(x_0)(x - x_0) = 0$$

for x so as to obtain $x_1 = x_0 - Df(x_0)^{-1} f(x_0)$, or the recursion formula

$$\begin{cases} x_0 \text{ given}, \\ x_{n+1} = x_n - Df(x_n)^{-1} f(x_n), \quad n \geqslant 1. \end{cases} \tag{2.3}$$

Notice that this is only possible as long as $Df(x_n)$ is invertible, which imposes restrictions on f as well as on the spaces V and W. If $V = \mathbb{R}^n$ and $W = \mathbb{R}^m$ for $m, n \in \mathbb{N}$, we would need to have $m = n$. This procedure is called *Newton's method* and can be proved to deliver a solution of $f(x) = 0$ "in the limit" (we are about to define this), provided f satisfies a regularity assumption (some differentiability) and x_0 is chosen close enough to an actual zero of f. For now, let us just interpret this as saying that x_n yields a better and better approximation of a zero as n gets larger. This is not the only available procedure to compute zeros of a function but it is widespread and one of the pillars of numerical analysis. In our concrete example above, we have $V, W = \mathbb{R}$, and $f(x) = x^2 - 2$ for $x \in \mathbb{R}$. Since $f'(x) = 2x$, we see that (2.2) is an example of (2.3) as

$$x_{n+1} = x_n - f'(x_n)^{-1} f(x_n) = x_n - \frac{x_n^2 - 2}{2x_n} = \frac{x_n}{2} + \frac{1}{x_n}.$$

2.3.3 Construction of the reals II

Returning to (2.2), we would like to understand the behavior of x_n for large n. In order to be able to say when two rational numbers are close to each other, we define a distance function $d_{\mathbb{Q}} : \mathbb{Q} \times \mathbb{Q} \to \mathbb{Q}^+$ by

$$d_{\mathbb{Q}}(x,y) = |x-y|, \quad x,y \in \mathbb{Q},$$

where $|\cdot| : \mathbb{Q} \to \mathbb{Q}$ is the *absolute value* function defined by

$$|x| = \begin{cases} x, & x \in \mathbb{Q}^+, \\ 0, & x = 0, \\ -x, & -x \in \mathbb{Q}^+. \end{cases}$$

For the function $d_{\mathbb{Q}}$, it holds that $d_{\mathbb{Q}}(x,y) = 0$ if and only if $x = y$, that $d_{\mathbb{Q}}(x,y) \geqslant 0$ for any $x, y \in \mathbb{Q}$, and that

$$d_{\mathbb{Q}}(x,z) \leqslant d_{\mathbb{Q}}(x,y) + d_{\mathbb{Q}}(y,z) \quad \forall x,y,z \in \mathbb{Q}.$$

The reader is encouraged to verify the validity of these claims (Q8). Following our intuition, we will say that x_n gets closer and closer to $x_\infty \in \mathbb{Q}$ if no matter how large $M \in \mathbb{N}$ is

$$x_n \in \mathbb{B}\Big(x_\infty, \frac{1}{M}\Big) = \Big(x_\infty - \frac{1}{M}, x_\infty + \frac{1}{M}\Big),$$

for all but finitely many $n \in \mathbb{N}$. In more mathematical terms, we say that x_n *converges to x_∞ as n goes to* ∞, for short $x_n \to x_\infty$ as $n \to \infty$, or more concisely, $\lim_{n\to\infty} x_n = x_\infty$ if

$$\forall M \in \mathbb{N}\ \exists N \in \mathbb{N} \text{ s. t. } d_{\mathbb{Q}}(x_n, x_\infty) = |x_n - x_\infty| \leqslant \frac{1}{M} \text{ for } n \geqslant N. \tag{2.4}$$

The set of rational numbers has the so-called Archimedean property, which states that any $q \in \mathbb{Q}$ satisfies $q < m$ for some $m \in \mathbb{N}$. If $q \leqslant 0$, this is clearly true, otherwise $q \in \mathbb{Q}^+$ and we can write $q = \frac{m}{n} (= [m,n])$ for $m, n \in \mathbb{N}$, so that $q = [m,n] \leqslant [m,1] = m$. This discussion shows that there is no positive rational number smaller than $\frac{1}{m}$ for all $m \in \mathbb{N}$ (why?) and, in turn, provides some intuition for the definition of convergence.

An example of a convergent sequence is obtained by setting $x_n = \frac{1}{n}$ for $n \in \mathbb{N}$. In this case, $x_\infty = 0$ and the sequence converges to 0. Notice that an arbitrary sequence $(x_n)_{n\in\mathbb{N}} \in \mathbb{Q}^{\mathbb{N}}$ does not need to converge as is exemplified by the sequences given by $x_n = (-1)^n$ or by $x_n = n$ for $n \in \mathbb{N}$. Convergence can, however, fail in more subtle ways. We will now show that the sequence $(x_n)_{n\in\mathbb{N}}$ defined in (2.2) does not converge in spite of the fact that

$$\lim_{n\to\infty} x_n^2 = 2.$$

First, notice that the recursive definition of $(x_n)_{n\in\mathbb{N}}$ ensures that $x_n \in \mathbb{Q}$, since the construction of the sequence only involves the operations of addition and multiplication (albeit by the reciprocal of a number) on account that $\mathbb{Q}$ is a field. Next, define the sequence $(y_n)_{n\in\mathbb{N}}$ by

$$\begin{cases} y_0 = x_0^2 \in \mathbb{Q}, \\ y_{n+1} = x_{n+1}^2 = (x_n/2 + 1/x_n)^2 = y_n/4 + 1 + 1/y_n, \quad n \geqslant 0, \end{cases}$$

which also consists of rational numbers. We claim that $y_n \to 2$ as $n \to \infty$. Observe that

$$y/4 + 1 + 1/y > 2 \Leftrightarrow (y-2)^2 > 0$$

implies that $y_1 > 2$ so long as $y_0 \neq 2$. If $y_0 = 2$, the sequence becomes stationary, i. e., $y_n = 2$ for each $n \in \mathbb{N}$. In particular, whatever $2 \neq y_0 > 0$ is, $y_1 > 2$, and thus by induction $y_n > 2$ for all $n \in \mathbb{N}$. Now if $y_n = 2 + \epsilon$ for some $\epsilon \in \mathbb{Q}^+$, then

$$y_{n+1} - 2 = \frac{16 + 8\epsilon + \varepsilon^2}{4(2+\epsilon)} - 2 = \frac{\epsilon^2}{4(2+\epsilon)},$$

and thus

$$|y_{n+1} - 2| = \frac{\epsilon}{4}\frac{\epsilon}{2+\epsilon} \leqslant \frac{|y_n - 2|}{4},$$

since $\frac{\varepsilon}{2+\varepsilon} \leqslant 1$. This implies that

$$|y_n - 2| \leqslant \frac{|y_0 - 2|}{4^n}, \qquad n \in \mathbb{N},$$

and convergence follows. From $y_n > 2 > 1$ for $n \geqslant 1$, we infer that

$$|x_n - x_m| = |\sqrt{y_n} - \sqrt{y_m}|\frac{\sqrt{y_n} + \sqrt{y_m}}{\sqrt{y_n} + \sqrt{y_m}} = \frac{|y_n - y_m|}{x_n + x_m} \leqslant \frac{1}{2}|y_n - y_m|,$$

for $n, m \geqslant 1$. As $|y_n - y_m|$ can be made arbitrarily small by upping the indices $m, n \geqslant 1$, the term $|x_n - x_m|$ can also be made arbitrarily small in the same way. More precisely, given $M \in \mathbb{N}$, there is $N \in \mathbb{N}$ such that

$$|x_n - x_m| \leqslant \frac{1}{M}, \qquad \text{whenever } n, m \geqslant N.$$

Check the validity of this for yourself. Now, this may trick us into believing that the sequence $(x_n)_{n\in\mathbb{N}}$ converges. However, if it did, that is, if $x_n \to x_\infty$ for some $x_\infty \in \mathbb{Q}$, then we would have to conclude that

$$x_n^2 = y_n \longrightarrow x_\infty^2 = y_\infty = 2 \quad \text{as } n \to \infty.$$

We do, however, know that the equation $x^2 = 2$ has no solution in $\mathbb{Q}$. We therefore need to reject the idea that $(x_n)_{n\in\mathbb{N}}$ does converge. This shows that it is possible for a sequence to satisfy the condition

$$\forall M \in \mathbb{N}\ \exists N \in \mathbb{N} \text{ s. t. } d_{\mathbb{Q}}(x_n, x_m) = |x_n - x_m| \leqslant \frac{1}{M} \text{ for } n, m \geqslant N, \tag{2.5}$$

and simultaneously fail to converge. When a sequence satisfies (2.5), it is said to be a *Cauchy sequence*. Prove (Q9) that any convergent sequence is a Cauchy sequence. The first few terms of the special sequence $(x_n)_{n\in\mathbb{N}}$ that we constructed are given by

$$x_1 = 1.8\bar{3}, \quad x_2 = \underline{1}.46\overline{21}, \quad x_3 = \underline{1.41}4998\cdots,$$
$$x_4 = \underline{1.41421}37\cdots, \quad x_5 = \underline{1.41421356}, \quad \ldots$$

if $x_0 = 3$. It appears as if more and more digits settle on a value as the sequence index grows. We can intuitively think that this specific Cauchy sequence is pointing to the “new” number $\sqrt{2}$ in spite of the fact that it does not converge. Since we do not have that number, the largest set of numbers we constructed so far is indeed $\mathbb{Q}$, we could take the whole sequence to “be” the number $\sqrt{2}$ since it “detects it”. Just like in the construction of whole and rational numbers, however, simply defining new numbers as Cauchy sequences would cause ambiguity. For any $x_0 > 0$, the recursion (2.2) yields a distinct sequence, which would need to be interpreted as $\sqrt{2}$. The issue can be avoided in a way similar to the one we used in the construction of the rationals (or of the whole numbers): build equivalence classes. First, define the set

$$\mathbb{Q}_{cs}^{\mathbb{N}} = \{x = (x_n)_{n\in\mathbb{N}} \in \mathbb{Q}^{\mathbb{N}} \mid x \text{ is a Cauchy sequence}\}$$

of all Cauchy sequences of rational numbers. Then introduce on this set the equivalence relation given by

$$(x_n)_{n\in\mathbb{N}} = x \sim y = (y_n)_{n\in\mathbb{N}} \Longleftrightarrow \lim_{n\to\infty}(x_n - y_n) = 0.$$

It makes sense to think of two sequences $x, y \in \mathbb{Q}_{cs}^{\mathbb{N}}$ as encoding the same number if their difference vanishes in the limit. When both sequences converge, the relation $x \sim y$ boils down to $\lim_{n\to\infty} x_n = \lim_{n\to\infty} y_n$, which justifies the definition. The set of *real numbers* is then defined by

$$\mathbb{R} = \mathbb{Q}_{cs}^{\mathbb{N}}/\sim,$$

i. e., as the set of equivalence classes $x = [x_1, x_2, \dots]$ of Cauchy sequences. In the case of whole numbers and of rationals, it was enough to consider pairs of naturals; here, we need infinite sequences of rationals to pin-point real numbers. From an intuitive point of view, this is to be expected since a real number can require infinitely many digits for its description, and as such, cannot be fully captured by any finite sequence. The need to resort to infinite sequences (and to convergence) is a hallmark of analysis and a feature that distinguishes it from algebra.

To make sure that we have a workable definition of real numbers, we need to verify that $\mathbb{Q}$ can be realized as a subset of $\mathbb{R}$ and that the operations of addition and multiplication as well as the concept of distance can be extended to the new number set. Given a rational number $q \in \mathbb{Q}$, we can associate to it the sequence $(q_n)_{n\in\mathbb{N}} = (q, q, \dots)$ and its whole equivalence class $[q, q, \dots]$. In order to be able to interpret

$$\iota : \mathbb{Q} \to \mathbb{R}, \quad q \mapsto [q, q, \dots],$$

as the inclusion $\mathbb{Q} \subset \mathbb{R}$, it needs to be shown injective. To this end, take sequences $(q_n)_{n\in\mathbb{N}} \in [q, q, \dots]$ and $(p_n)_{n\in\mathbb{N}} \in [p, p, \dots]$ for $q \neq p$. Then we have that

$$\lim_{n\to\infty} (q_n - q) = 0 \quad \text{and} \quad \lim_{n\to\infty} (p_n - p) = 0,$$

and, consequently, that

$$\lim_{n\to\infty} (q_n - p_n) = q - p \neq 0,$$

so that $[(q_n)_{n\in\mathbb{N}}] \neq [(p_n)_{n\in\mathbb{N}}]$ or $[q, q, \dots] \cap [p, p, \dots] = \emptyset$. If $q \in \mathbb{Q}$, we simply write q for the real number $[q, q, \dots]$. Notice that

$$[q_1, q_2, \dots] = q \in \mathbb{Q} \quad \text{whenever } \lim_{n\to\infty} q_n = q$$

for all $(q_n)_{n\in\mathbb{N}} \in [q_1, q_2, \dots]$ as already implicitly used above. We define addition and multiplication by

$$[q_1, q_2, \dots] + [p_1, p_2, \dots] = [q_1 + p_1, q_2 + p_2, \dots],$$
$$[q_1, q_2, \dots] \cdot [p_1, p_2, \dots] = [q_1 p_1, q_2 p_2, \dots],$$

respectively. For these to be acceptable definitions, it needs to be verified that they do not depend on the choice of class representatives. Let us begin with addition: take representatives $(\tilde{q}_n)_{n\in\mathbb{N}}, (\bar{q}_n)_{n\in\mathbb{N}} \in [q_1, q_2, \dots]$ as well as $(\tilde{p}_n)_{n\in\mathbb{N}}, (\bar{p}_n)_{n\in\mathbb{N}} \in [p_1, p_2, \dots]$. It follows that

$$\lim_{n\to\infty} (\tilde{q}_n - \bar{q}_n) = 0 \quad \text{and} \quad \lim_{n\to\infty} (\tilde{p}_n - \bar{p}_n) = 0,$$

which imply that

$$\lim_{n\to\infty}[(\tilde{q}_n + \tilde{p}_n) - (\bar{q}_n + \bar{p}_n)] = 0,$$

and hence that indeed $(\tilde{q}_n + \tilde{p}_n)_{n\in\mathbb{N}} \sim (\bar{q}_n + \bar{p}_n)_{n\in\mathbb{N}}$. The reader is urged to verify by applying the definition of convergence for sequences in $\mathbb{Q}^{\mathbb{N}}$ that indeed

$$\lim_{n\to\infty}(q_n + p_n) = \lim_{n\to\infty} q_n + \lim_{n\to\infty} p_n,$$

for any two convergent sequences $(q_n)_{n\in\mathbb{N}}$ and $(p_n)_{n\in\mathbb{N}}$ of rationals. This fact was used in the above argument. As for multiplication, an additional observation is needed. It holds that any Cauchy sequence $(q_n)_{n\in\mathbb{N}}$ must necessarily be bounded if, by *bounded*, we mean that

$$|q_n| \leqslant M \quad \text{for } n \in \mathbb{N},$$

for some constant $M \in \mathbb{N}$ (a bound). The Cauchy property means that, given $\varepsilon \in \mathbb{Q}^+$, $N = N(\varepsilon) \in \mathbb{N}$ can be found such that

$$|q_n - q_m| \leqslant \varepsilon \quad \text{for } n, m \geqslant N(\varepsilon).$$

Setting $m = N$ and disentangling the absolute value yields

$$q_N - \varepsilon \leqslant q_n \leqslant q_N + \varepsilon \quad \text{for } n \geqslant N(\varepsilon).$$

It now suffices to choose $\varepsilon = 1$ to see that

$$|q_n| \leqslant \max\{|q_{N(1)} - 1|, |q_{N(1)} + 1|, \max\{|q_1|, |q_2|, \dots, |q_{N(1)-1}|\}\} = M,$$

which entails boundedness with bound M. Returning to the definition of product: take any representative sequences $(\tilde{q}_n)_{n\in\mathbb{N}}, (\bar{q}_n)_{n\in\mathbb{N}} \in [q_1, q_2, \dots]$ as well as $(\tilde{p}_n)_{n\in\mathbb{N}}, (\bar{p}_n)_{n\in\mathbb{N}} \in [p_1, p_2, \dots]$. These are all bounded sequences and we can assume that they share a common bound M (the maximum of any of their individual ones). We can therefore verify that

$$\begin{aligned} |\tilde{q}_n \cdot \tilde{p}_n - \bar{q}_n \cdot \bar{p}_n| &\leqslant |\tilde{q}_n \cdot (\tilde{p}_n - \bar{p}_n) + (\tilde{q}_n - \bar{q}_n) \cdot \bar{p}_n| \\ &\leqslant |\tilde{q}_n||\tilde{p}_n - \bar{p}_n| + |\tilde{q}_n - \bar{q}_n||\bar{p}_n| \\ &\leqslant M(|\tilde{p}_n - \bar{p}_n| + |\tilde{q}_n - \bar{q}_n|). \end{aligned}$$

Now, by the definition of the equivalence relation, it is possible to make the term on the right-hand side as small as one wishes by choosing n large enough to yield that

$$|\tilde{q}_n \cdot \tilde{p}_n - \bar{q}_n \cdot \bar{p}_n| \leqslant \varepsilon \quad \text{for } n \geqslant N(\varepsilon).$$

We just choose $N_1(\varepsilon), N_2(\varepsilon) \in \mathbb{N}$ such that

$$|\tilde{q}_n - \bar{q}_n| \leqslant \frac{\varepsilon}{2M} \text{ for } n \geqslant N_1(\varepsilon) \quad \text{and} \quad |\tilde{p}_n - \bar{p}_n| \leqslant \frac{\varepsilon}{2M} \text{ for } n \geqslant N_2(\varepsilon),$$

and set $N(\varepsilon) = \max\{N_1(\varepsilon), N_2(\varepsilon)\}$ for any arbitrary chosen $\varepsilon > 0$. This shows that $[\tilde{q}_1\tilde{p}_1, \tilde{q}_2\tilde{p}_2, \ldots] \sim [\bar{q}_1\bar{p}_1, \bar{q}_2\bar{p}_2 \ldots]$ whenever $(\tilde{q}_n)_{n\in\mathbb{N}} \sim (\bar{q}_n)_{n\in\mathbb{N}}$ and $(\tilde{p}_n)_{n\in\mathbb{N}} \sim (\bar{p}_n)_{n\in\mathbb{N}}$, as desired.

It can (and should) also be verified that $\iota(q+p) = \iota(p)+\iota(q)$ and that $\iota(q \cdot p) = \iota(p) \cdot \iota(q)$ for any $q, p \in \mathbb{Q}$. This shows that the newly defined operations do indeed extend the ones for rationals. It is left as an exercise to check that addition and multiplication of reals enjoy all the properties of these operations for the rationals like commutativity, associativity, existence of an identity element and of additive and multiplicative inverses, and distributivity of the two operations. The set $\mathbb{Q}$ is therefore a field as are the sets $\mathbb{R}$ and the soon to be introduced $\mathbb{C}$.

The irrational number denoted by $\sqrt{2}$ was introduced as the equivalence class of the sequence $x = (x_n)_{n\geqslant 0}$ recursively defined by (2.2) and any initial value $x_0 \in \mathbb{Q}^+$. With the newly defined multiplication for reals, we now see that

$$\begin{aligned} x^2 &= x \cdot x = [x_0, x_1, x_2, \ldots] \cdot [x_0, x_1, x_2, \ldots] \\ &= [x_0^2, x_1^2, x_2^2, \ldots] = [y_0, y_1, y_2, \ldots] \\ &= 2, \end{aligned}$$

since $\lim_{n\to\infty} y_n = 2$, thus providing a justification for its name $\sqrt{2}$.

Starting with the set $\mathbb{Q}$ of rationals and with a way to measure distance between numbers (points), we introduced the concept of convergence to model our intuition of a sequence of points getting arbitrarily close to a (limit) point as its index grows. We learned that there exist sequences of points, which come arbitrarily close to each other as the index grows but still do not converge. In more mathematical terms, we discovered that not all Cauchy sequences do necessarily converge. When they do not, the example $(x_n)_{n\in\mathbb{N}}$ we considered in (2.2) suggests that they detect "holes" in the set considered, like the sequence $(x_n)_{n\in\mathbb{N}}$ seems to "feel" the presence of the number $\sqrt{2}$. We therefore took these sequences (equivalence classes thereof) and stipulated that they be the "new" numbers themselves, hence effectively filling in the holes they helped us discover.

2.3.4 Metric spaces and completeness

The above procedure, the construction of $\mathbb{R}$ from $\mathbb{Q}$ by means of Cauchy sequences, is a simple manifestation of a general approach that proves extremely useful in analysis. One can in fact begin with any set M, which admits a distance d_M between any two of its points and define the concepts of convergence and of Cauchy sequence. Then one verifies whether there are any Cauchy sequences that do not converge. If that is not the case, the space (M, d_M), which is typically referred to as a *metric space*, is called *complete*. Intuitively, this means that it contains no holes. If, on the other hand, it is not complete, it can be extended to a complete space by a procedure analogous to the one used above, i. e., by considering the set of equivalence classes of Cauchy sequences. The procedure goes by the name of *completion* of the space (M, d_M). We shall consider a more involved concrete example in Chapter 3. Modeling a distance function on the absolute value that was used for the rationals leads to the concept of *metric* d_M on a (nonempty) set M. It is a function $d_M : M \times M \to \mathbb{R}$ satisfying the following properties:

(m1) $d_M(x, y) \geqslant 0$ for $x, y \in M$, with equality only if $x = y$.
(m2) For any $x, y \in M$, it holds that $d_M(x, y) = d_M(y, x)$.
(m3) The *triangle inequality* holds, i. e., $d_M(x, z) \leqslant d_M(x, y) + d_M(y, z)$ for any $x, y, z \in M$.

Notice that condition **(m3)** encodes the idea that any path realizing the distance from the point x to the point z should be shorter than a path taking one first from x to another point y, and from there to z. Think of three points X, Y, Z in a Euclidean plane in general position, and hence forming a triangle: going straight from X to Z, you should cover less ground than going through Y first. What is the intuitive expectation encoded by the second condition? Can you envision practical situations where you would consider it inappropriate? On the set $\mathbb{Q}$, we use the distance given by $d_{\mathbb{Q}}(q, p) = |q - p|$, which makes use of addition and of the absolute value function. On a general set M, where no addition may be available, one has to settle for the more general definition above. How would you define a distance on the set of reals? See below for an answer and compare it to yours.

With a metric in hand, we can define what it means for a sequence $(x_n)_{n \in \mathbb{N}} \in M^{\mathbb{N}}$ to *converge* to a limit $x_\infty \in M$ as the validity of

$$\forall \varepsilon > 0 \; \exists N \in \mathbb{N} \text{ s. t. } d_M(x_n, x_\infty) \leqslant \varepsilon \text{ for } n \geqslant N. \tag{2.6}$$

More concisely, we can say that, given any $\varepsilon > 0$, at most finitely many elements of the sequence are at a distance to x_∞, which is larger than ε. A Cauchy sequence, in this context, would amount to a sequence satisfying

$$\forall \varepsilon > 0 \; \exists N \in \mathbb{N} \text{ s. t. } d_M(x_n, x_m) \leqslant \varepsilon \text{ for } m, n \geqslant N. \tag{2.7}$$

A visual example of a metric space is the surface of the earth (as the set) and the length of the shortest "path" connecting any two points on it as the distance function/metric.

2.3.5 Construction of the reals III

Returning to the reals and, after defining addition and multiplication for them, we would like to be able to introduce an order as well like we have for rationals, and a distance function that extends the one for rationals. The discussion will be concluded by proving that the set $\mathbb{R}$, in contrast to $\mathbb{Q}$, is now complete. In order to obtain an order on $\mathbb{R}$, we need to be able to say when an equivalence class of Cauchy sequences is nonnegative. We stipulate that the real number $x = [x_1, x_2, \ldots]$ is positive, i. e., $x > 0$, iff there is $M \in \mathbb{N}$ such that

$$\forall (x_n)_{n\in\mathbb{N}} \in x \; \exists N \in \mathbb{N} \text{ s. t. } x_n \geqslant \frac{1}{M} > 0 \text{ for } n \geqslant N.$$

The reader is encouraged to verify (Q10) that it is enough to establish the validity of this condition for a single representative of x. Of course, we would write $x \geqslant 0$ for $x \in \mathbb{R}$ iff $x = 0$ or $x > 0$ as well as $x \leqslant y$ for $x, y \in \mathbb{R}$ iff $y - x \geqslant 0$. We can now define an absolute value for reals by

$$|x|_{\mathbb{R}} = \begin{cases} x, & 0 \leqslant x \in \mathbb{R}, \\ -x, & 0 > x \in \mathbb{R}, \end{cases}$$

and with it a distance function for reals in the same way as we did for rationals, $d_{\mathbb{R}}(x, y) = |x - y|_{\mathbb{R}}$ for $x, y \in \mathbb{R}$. Verify that positivity and absolute value do indeed extend the corresponding concepts for rationals and show that

$$|x|_{\mathbb{R}} = \big|[x_1, x_2, \ldots]\big|_{\mathbb{R}} = [|x_1|, |x_2|, \ldots], \quad x \in \mathbb{R}.$$

As for convergence for sequences of real numbers, the definition can be made via (2.6) using $d_{\mathbb{R}}$. In view of the way the reals are constructed, it should not come as a surprise that $\mathbb{R}$ is now complete, i. e., it does not admit non-convergent Cauchy sequences. We effectively filled the "holes" we had found in $\mathbb{Q}$. In order to give a formal proof, we first observe that, given any $x \in \mathbb{R}$ and any $M \in \mathbb{N}$, a rational number q can be found such that

$$|q - x|_{\mathbb{R}} \leqslant \frac{1}{M}.$$

This means that a real number admits arbitrarily close rational numbers.[2] To see this, let $x = [q_1, q_2, \ldots]$ and consider any representative $(\tilde{q}_n)_{n\in\mathbb{N}} \in x$. As the latter is a Cauchy sequence, we can find $N \in \mathbb{N}$ such that $|\tilde{q}_n - \tilde{q}_m| \leqslant \frac{1}{M}$ provided $m, n \geqslant N$. It follows that

$$\big|[\tilde{q}_1, \tilde{q}_2, \ldots] - [\tilde{q}_N, \tilde{q}_N, \ldots]\big|_{\mathbb{R}} = [|\tilde{q}_1 - \tilde{q}_N|, |\tilde{q}_2 - \tilde{q}_N|, \ldots] \leqslant \frac{1}{M}.$$

To conclude the proof of completeness, we need to take an arbitrary Cauchy sequence $(x_n)_{n\in\mathbb{N}} \in \mathbb{R}^{\mathbb{N}}$ of reals and show that it possesses a limit. For $n \in \mathbb{N}$, we pick $q_n \in \mathbb{Q}$

2 One says that $\mathbb{Q}$ is *dense* in $\mathbb{R}$.

with $|x_n - q_n| \leqslant \frac{1}{n}$. This yields a real number $x = [q_1, q_2, \ldots]$ since it can be shown (do it (Q11)) that $(q_n)_{n\in\mathbb{N}}$ is a Cauchy sequence (using that $(x_n)_{n\in\mathbb{N}}$ is one). Finally, it holds that $\lim_{n\to\infty} x_n = x$, since, given $M \in \mathbb{N}$, it is always possible to find $N \in \mathbb{N}$ such that

$$|x_n - q_n|_{\mathbb{R}} \leqslant \frac{1}{M} \quad \text{for } n \geqslant N.$$

If this feels like "cheating" to you, it actually is. The only crucial observation is really that there are sequences of rationals, which are Cauchy but do not converge. In order to make them converge, we use them as the object they are telling us is missing (the hole). This, just as it was the case for whole numbers and rationals, can only be done by resorting to equivalence classes to remove ambiguity. In the case of reals, however, it is also necessary to consider infinite sequences and not simple pairs. While we accept the new number system, forget about its actual nature (equivalence classes of Cauchy sequences of rationals) and operate with it using only the properties it satisfies, we should not lose sight of the fact that a real number cannot possibly be captured by any finite number of field operations (unless it is rational, of course). It requires an infinite number of digits to describe, after all. More precisely, a real number does not have a finite expansion in any integer basis (as it would be rational in that case), while rational numbers always do. The number 2/3, for instance, can be written as 0.2 in base 3.[3]

2.4 Complex numbers

Real numbers are often motivated algebraically by saying that they are needed to obtain a solution to equations like $x^2 = 2$ or more in general to polynomial equations. If the latter is the ultimate goal, then one quickly encounters equations, which do not admit real solutions, such as $x^2 = -1$. Complex numbers are often justified by the attempt to obtain a number of solutions to polynomial equations that matches the polynomial's degree. In this sense, complex numbers conclude the quest for a so-called algebraically closed field. Here, we choose a somewhat similar but more analytical justification of the need for complex numbers. We start by a simple but useful algebraic identity

$$(1 - y)(1 + y + y^2 + \cdots + y^n) = 1 - y^{n+1},$$

which is valid for any $y \in \mathbb{R}$ and $n \in \mathbb{N}$. In particular, we can choose $y = -x^2$ for some $x \in \mathbb{R}$ and obtain

$$\frac{1 - (-x^2)^{n+1}}{1 + x^2} = 1 - x^2 + x^4 \mp \cdots + (-1)^n x^{2n}.$$

---

3 As a matter of a fact, most real numbers are even transcendental, i. e., not algebraic, in the sense that they are not zeros of polynomials with rational coefficients. In fact, algebraic numbers such as $\sqrt{2}$ are countable.

Next, we consider the function $f(x) = \frac{1}{1+x^2}$ for $x \in \mathbb{R}$, which is a very nice function: it is bounded by 1, positive, and has infinitely many derivatives. Taking n to ∞ in the above identity yields that

$$\frac{1}{1+x^2} = \sum_{k=0}^{\infty} (-1)^k x^{2k},$$

when $|x| < 1$. In this way, we obtain a series representation for the function

$$f(x) = \frac{1}{1+x^2}, \quad x \in \mathbb{R}.$$

While the function does not experience any "issue" as $|x|$ becomes larger than 1, this series representation stops converging for $|x| \geqslant 1$. It is natural to ask why that is. In order to obtain an answer, however, we need to see beyond real numbers. The reason does, in fact, originate in the presence of zeros of the denominator $1+x^2$, which happen to be of size one. This leads to the construction of complex (or imaginary) numbers. The procedure is purely algebraic. Consider pairs $(x, y) \in \mathbb{R}^2$ of real numbers and define the operations on the set of pairs given by

$$(x, y) + (\tilde{x}, \tilde{y}) = (x + \tilde{x}, y + \tilde{y}) \quad \text{and} \quad (x, y) \cdot (\tilde{x}, \tilde{y}) = (x\tilde{x} - y\tilde{y}, x\tilde{y} + \tilde{x}y)$$

for $x, \tilde{x}, y, \tilde{y} \in \mathbb{R}$. We can think of $\mathbb{R}$ as a subset by the identification

$$\iota : \mathbb{R} \to \mathbb{R}^2, \quad x \mapsto (x, 0)$$

and notice that the new addition and multiplication extend the corresponding operations in $\mathbb{R}$ since

$$\iota(x + \tilde{x}) = (x + \tilde{x}, 0) = (x, 0) + (\tilde{x}, 0) = \iota(x) + \iota(\tilde{x}),$$

and

$$\iota(x\tilde{x}) = (x\tilde{x}, 0) = (x, 0) \cdot (\tilde{x}, 0) = \iota(x) \cdot \iota(\tilde{x}).$$

It turns out that $(\mathbb{R}^2, +, \cdot)$ inherits all algebraic properties of $(\mathbb{R}, +, \cdot)$. Check the validity of all the properties of addition and multiplication that make this set a field. The numbers $(0, \pm 1)$ of this new number system satisfy

$$(0, \pm 1)^2 = (0, \pm 1) \cdot (0, \pm 1) = (-1, 0) = -1,$$

so that the equation $x^2 = -1$ now has the two solutions $x = (0, -1)$ and $x = (0, 1)$. In fact, even more is true: it can be proved that any polynomial equation

$$x^n + a_{n-1}x^{n-1} + \cdots + a_1 x + a_0 = 0$$

with coefficients $a_0, a_1, \dots, a_{n-1} \in \mathbb{C}$ admits n solutions in this new set (counting their multiplicity). This is known as the fundamental theorem of algebra.

The elements of the new number field are called *complex (or imaginary) numbers*. It is customary to write $z = (x, y)$ simply as $z = x + iy$, where x is called *real part* and y *imaginary part* of the number z. The collection of all complex numbers is denoted by $\mathbb{C}$. Noticing that $i^2 = (0,1)^2 = -1$, this notation is justified by the fact that

$$\begin{aligned}(x+iy)(u+iv) &= xu + i^2yv + i(xv+yu)\\ &= xu - yv + i(xv+yu) \quad \forall x,y,u,v \in \mathbb{R}\end{aligned}$$

formally applying commutativity of multiplication and distributivity. The absolute value for real numbers is extended to the *modulus* of complex numbers given by

$$|z|_{\mathbb{C}} = \sqrt{x^2+y^2} = \sqrt{(x+iy)(x-iy)} = \sqrt{z\bar{z}} \quad \text{for } z = x+iy,$$

where $\bar{z} = x - iy \in \mathbb{C}$ denotes the so-called *complex conjugate* of the number $z = x + iy \in \mathbb{C}$. Observe that $|x|_{\mathbb{R}} = |x + i0|_{\mathbb{C}}$. We conclude this discussion by observing that the function $f(x) = \frac{1}{1+x^2}$ can be extended to all complex numbers z simply by setting

$$f(z) = \frac{1}{1+z^2} = \frac{1}{1+(x+iy)^2} = \frac{1}{1+x^2-y^2+i2xy}, \quad z \in \mathbb{C}.$$

It still holds that

$$f(z) = \sum_{k=0}^{\infty} (-1)^k z^{2k}$$

for $|z|_{\mathbb{C}} < 1$ and we now know that the reason for the breakdown of this identity is division by zero, which occurs for $z = \pm i$. While the problem manifested itself on the real line already, its roots (pun intended) are to be traced back to the existence of complex zeros of the denominator.

3 Vectors and their spaces

In virtually all branches of mathematics, the concept of vector plays a central role. It is the mathematical object encoding the intuition of direction. We often think of vectors as directed segments, which can be added by juxtaposition/concatenation as depicted below and which can be stretched by any factor.

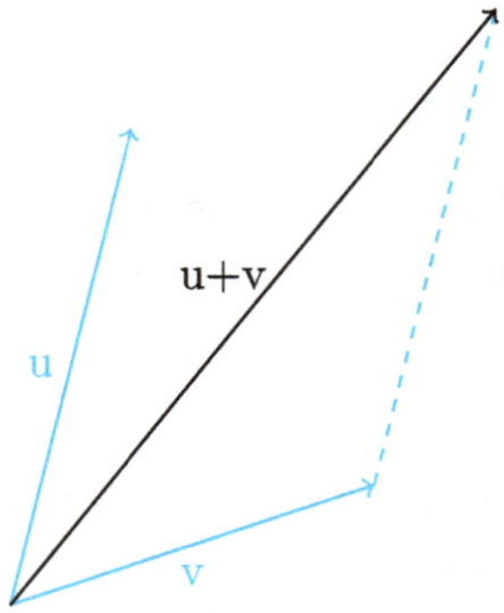

The mathematical structure that captures vectors is that of *vector space*. It consists of a set, the elements of which can be added and stretched. It is therefore necessary that it be possible to define these operations of addition and (scalar) multiplication satisfying some properties suggested by one's intuition or by applications that motivate their introduction (e. g., physics). Scalar multiplication is used to stretch vectors and one needs to choose a field of numbers $\mathbb{F}$ like $\mathbb{Q}$, $\mathbb{R}$, $\mathbb{C}$, or any other (possibly finite) field for the stretching factors.

We define a triple $(V, +, \cdot)$ comprised of a set V and maps $+ : V \times V \to V$ and $\cdot : \mathbb{F} \times V \to V$ to be a *vector space over the field* $\mathbb{F}$ if the following properties are satisfied:

(vs1) V contains at least one element, called the zero vector 0 for which it holds that $v + 0 = 0 + v = v$ for all $v \in V$.

(vs2) Addition is commutative, i. e., $u + v = v + u$ for any $u, v \in V$.

(vs3) Addition is associative, i. e., $(u + v) + w = u + (v + w)$ for any $u, v, w \in V$.

(vs4) Each nonzero vector u (if any) has an inverse vector $-u$, for which it holds that $u + (-u) = (-u) + u = 0$.

(vs5) There is compatibility between the multiplication in the field $\mathbb{F}$ and scalar multiplication of vectors, i. e., it holds that $\alpha \cdot (\beta \cdot u) = (\alpha\beta) \cdot u$ for any $\alpha, \beta \in \mathbb{F}$ and $u \in V$.

(vs6) The multiplicative identity 1 of the field $\mathbb{F}$ acts like an identity for scalar multiplication, i. e., it holds that $1 \cdot u = u$ for any $u \in V$.

(vs7) Distributivity in both variables holds, i. e., we have that $(\alpha + \beta) \cdot u = \alpha \cdot u + \beta \cdot u$ and $\alpha \cdot (u + v) = \alpha \cdot u + \alpha \cdot v$ for all $\alpha, \beta \in \mathbb{F}$ and all $u, v \in V$.

https://doi.org/10.1515/9783110780925-003

Notice how $(V, +)$ is a commutative group according to the definition of the previous chapter. As an exercise verify that the inverse $-u$ of a vector $u \in V$ is unique, that $0 = 0_{\mathbb{F}} \cdot u$ (where the index of the second 0 is used to stress that it is the zero element/additive identity of the field $\mathbb{F}$), and that $-u = (-1) \cdot u$, where -1 is the additive inverse of 1 in the field $\mathbb{F}$. It is understood that it is only allowed to use properties (**vs1**)–(**vs7**) in the verification. The simplest examples of vector spaces are the trivial vector space $V = \{0\}$ consisting of a single vector and the field $\mathbb{F}$ itself, which are easily seen to be vector spaces for any arbitrary field $\mathbb{F}$. Given any set S and a field $\mathbb{F}$, the set $\mathbb{F}^S$ of all functions from S to $\mathbb{F}$ has the structure of a vector space over $\mathbb{F}$. Addition and scalar multiplication are given by

$$+ : \mathbb{F}^S \times \mathbb{F}^S \to \mathbb{F}^S,\ (u, v) \mapsto +(u, v) \equiv u + v : S \to \mathbb{F},\ s \mapsto u(s) + v(s)$$

$$\cdot : \mathbb{F} \times \mathbb{F}^S,\ (\alpha, u) \mapsto \cdot(\alpha, u) \equiv \alpha u : S \to \mathbb{F},\ s \mapsto \alpha u(s),$$

i. e., functions are added by adding their values and multiplied by multiplying their values, which is possible since the values are in the field $\mathbb{F}$. By selecting special sets S, we obtain a variety of examples. If

$$S = \mathbb{N}_n = \{1, 2, \dots, n\}$$

for $n \in \mathbb{N}$, the vector space $\mathbb{F}^{\mathbb{N}_n}$ is typically denoted by $\mathbb{F}^n$ and its elements are n-tuples $u = (u_1, \dots, u_n)$ with components $u_j \in \mathbb{F}$ for $j = 1, \dots, n$. Recall that we simplify the more precise notation $u(k)$ to u_k for $k \in \mathbb{N}_n$. The familiar vector spaces $\mathbb{R}^n$ and $\mathbb{C}^n$ are obtained in this way. If $S = \mathbb{N}$, one obtains the vector space $\mathbb{F}^{\mathbb{N}}$ of sequences in $\mathbb{F}$ with elements denoted by $(u_k)_{k \in \mathbb{N}}$ instead of the more accurate $u : \mathbb{N} \to \mathbb{F},\ k \mapsto u(k)$. Finally, by choosing $S = \mathbb{G} = \mathbb{G}^1$, or more in general, $S = \mathbb{G}^n$ for a field $\mathbb{G}$ and $n \in \mathbb{N}$, we obtain the vector space $\mathbb{F}^{\mathbb{G}^n}$ of functions $u : \mathbb{G}^n \to \mathbb{F}$. Selecting special fields yields the space $\mathbb{R}^{\mathbb{R}}$ of real real-valued functions $f : \mathbb{R} \to \mathbb{R}$, the space $\mathbb{C}^{\mathbb{R}}$ of complex-valued real functions $f : \mathbb{R} \to \mathbb{C}$, or the space $(\mathbb{R}^n)^{\mathbb{R}^m}$ of vector-valued functions $f : \mathbb{R}^m \to \mathbb{R}^n$ for $m, n \in \mathbb{N}$ are obtained.

In a vector space V, one can often find special subsets $W \subset V$ characterized by the fact that addition of elements of W and scalar multiplication do not produce outcomes outside W. Such subsets are called *vector subspaces* or *subspaces* and are more formally defined by the validity of the conditions

$$u + v \in W \quad \text{for any } u, v \in W,$$

$$\alpha u \in W \quad \text{for any } \alpha \in \mathbb{F}, u \in W.$$

Show that $(W, +|_{W \times W}, \cdot|_{\mathbb{F} \times W})$ is a vector space over $\mathbb{F}$ in its own right, where addition and scalar multiplication are defined by restriction and are well-defined by the validity of the above conditions. A simple example of a subspace is obtained starting with a vector $u \in V \setminus \{0\}$ and defining

$$\mathbb{F}u = \{\alpha u \mid \alpha \in \mathbb{F}\}.$$

Check that this is indeed a subspace, the subspace generated by the vector u, which is the smallest subspace containing u. In the vector space $\mathbb{F}^n$ and given a set of indices $\{i_1, i_2, \dots, i_m\}$ for $1 \leqslant m < n$, one can consider special subspaces of the form

$$V_{i_1,i_2,\dots,i_m} = \{u \in \mathbb{F}^n \mid u_{i_j} = 0 \text{ for } j = 1, \dots, m\},$$

such as, for instance, $\mathbb{R}^3 \supset V_2 = \mathbb{R} \times \{0\} \times \mathbb{R} \ni (u_1, 0, u_3)$ for $u_1, u_3 \in \mathbb{R}$. In the vector space $\mathbb{C}^{\mathbb{N}}$ of complex sequences, one can consider the sets

$$\begin{aligned}
\mathbb{C}^{\mathbb{N}}_{cs} &= \{u \in \mathbb{C}^{\mathbb{N}} \mid u \text{ is a Cauchy sequence}\},\\
\mathbb{C}^{\mathbb{N}}_{c} &= \{u \in \mathbb{C}^{\mathbb{N}} \mid u \text{ is convergent}\},\\
\mathbb{C}^{\mathbb{N}}_{b} &= \{u \in \mathbb{C}^{\mathbb{N}} \mid u \text{ is bounded}\},\\
\mathbb{C}^{\mathbb{N}}_{0} &= \left\{u \in \mathbb{C}^{\mathbb{N}} \mid \lim_{n\to\infty} u_n = 0\right\},
\end{aligned}$$

and show (exercise (Q1)) that they are subspaces. In the last example, would you still have a subspace if you replaced the limit 0 by another limit $0 \neq u_\infty \in \mathbb{C}$? Similarly, one can consider a variety of subspaces of the vector space $\mathbb{R}^{[0,1]}$ of real real-valued functions defined on the interval $[0, 1]$ such as

$$\begin{aligned}
\mathbb{R}^{[0,1]}_{b} &= \{u : [0,1] \to \mathbb{R} \mid u \text{ is bounded}\},\\
\mathbb{R}^{[0,1]}_{c} &= \{u : [0,1] \to \mathbb{R} \mid u \text{ is continuous}\},\\
\mathbb{R}^{[0,1]}_{0} &= \{u : [0,1] \to \mathbb{R} \mid u(0) = u(1) = 0\},\\
\mathbb{R}^{[0,1]}_{r} &= \{u : [0,1] \to \mathbb{R} \mid u \text{ is Riemann integrable}\}.
\end{aligned}$$

Think about what exactly you would need to show in order to verify that the above are all subspaces and try to carry out the detailed verification (if you are familiar with the required concepts of boundedness, continuity, and Riemann integrability).

3.1 Linear independence and bases

From now on, we will not be explicitly mentioning the base field when talking about a generic abstract vector space. We will do so, of course, if considering specific examples. Given n vectors $u_1, u_2, \dots, u_n$ in a vector space V, we can use them to generate more vectors by addition and scalar multiplication as in

$$\alpha_1 u_1 + \alpha_2 u_2 + \cdots + \alpha_n u_n = \sum_{k=1}^{n} \alpha_k u_k,$$

which is called a *linear combination* of the vectors $u_1, \ldots, u_n$ with *coefficients* $\alpha_1, \ldots, \alpha_n \in \mathbb{F}$. Show that, given $u_1, u_2, \ldots, u_n \in V$,

$$\left\{ \sum_{k=1}^{n} \alpha_k u_k \,\middle|\, \alpha_k \in \mathbb{F} \text{ for } k = 1, \ldots, n \right\} = \operatorname{span}\{u_1, \ldots, u_n\}$$

is a subspace of V, the smallest subspace containing all vectors $u_1, u_2, \ldots, u_n$ and called the *span* of the vectors $u_1, \ldots, u_n$. This gives us the idea that it may be possible to describe any vector in a vector space by using linear combinations of a "small" number of selected vectors. In order to keep the number of such vectors small, we need to make sure that they all genuinely contribute to the generation of additional vectors. This leads to the concept of *linear independence* which, for given vectors $u_1, u_2, \ldots, u_n$ amounts to requiring that the zero vector can only be described by the trivial linear combination of the given vectors, i. e., that

$$\sum_{k=1}^{n} \alpha_k u_k = 0 \quad \text{implies} \quad \alpha_1 = \cdots = \alpha_n = 0.$$

Notice that this is the same as saying that the map

$$C_{u_1,\ldots,u_n} : \mathbb{F}^n \to V, \quad \alpha = (\alpha_1, \ldots, \alpha_n) \mapsto \sum_{k=1}^{n} \alpha_k u_k,$$

is one-to-one. Indeed, notice that

$$\sum_{k=1}^{n} \alpha_k u_k = \sum_{k=1}^{n} \beta_k u_k,$$

is equivalent to

$$\sum_{k=1}^{n} (\alpha_k - \beta_k) u_k = 0,$$

from which we infer that $\alpha_k - \beta_k = 0$ for all $k = 1, \ldots, n$, and thus that $\alpha_k = \beta_k$ for all $k = 1, \ldots, n$, yielding injectivity. Now, given a vector space V, we either have $V = \{0\}$ and we can describe all its vectors with the vector $u = 0$ or we have $V \neq \{0\}$, in which case there must be a vector $u_1 \neq 0$. We are presented again with two alternatives: either $V = \mathbb{F}u_1$ or there is a vector $0 \neq u_2 \neq \alpha u_1$ $\forall \alpha \in \mathbb{F}$, i. e., satisfying $u_2 \notin \mathbb{F}u_1$. In the first case, we are done since the vector u_1 is enough to describe the whole vector space. In the latter case, we either have that

$$V = \mathbb{F}u_1 \oplus \mathbb{F}u_2 = \{\alpha_1 u_1 + \alpha_2 u_2 \,|\, \alpha_1, \alpha_2 \in \mathbb{F}\}$$

or we can continue the procedure by finding $V \ni u_3 \notin \operatorname{span}\{u_1, u_2\}$. Step-by-step, we either keep finding new linearly independent vectors indefinitely or the procedure stops after identifying $n \in \mathbb{N}$ vectors $u_1, \dots, u_n \in V$ such that any other vector $u \in V$ can be described as

$$u = \sum_{k=1}^{n} \alpha_k u_k$$

for scalars $\alpha_1, \dots, \alpha_n \in \mathbb{F}$, which are necessarily unique by construction. For this choice of vectors $u_1, \dots, u_n$, the map $C_{u_1,\dots,u_n} : \mathbb{F}^n \to V$ is both one-to-one and onto, i. e., bijective. The sequence $u_1, \dots, u_n$ is called a *basis* in this case and n, which can be shown to be uniquely determined,[1] is called the *dimension* of the vector space V. It can, however, happen that we can indefinitely add new linearly independent vectors without ever “filling out” V. In this case, the vector space is said to be infinite-dimensional. Taking $n \in \mathbb{N}$ and $\mathbb{F} = \mathbb{R}$, the vector space $\mathbb{R}^n$ has dimension n and the sequences $e_1, \dots, e_n$ and $f_1, \dots, f_n$ given by

$$(1, 0, \dots, 0), (0, 1, 0, \dots, 0), (0, 0, 1, 0, \dots, 0), \dots, (0, \dots, 0, 1) \quad \text{and}$$
$$(1, 0, \dots, 0), (1, 1, 0, \dots, 0), (1, 1, 1, 0, \dots, 0), \dots, (1, \dots, 1),$$

respectively, are two bases for it. For any $n \in \mathbb{N}$, the space $\mathbb{C}^n$ has dimension n over $\mathbb{C}$ and dimension $2n$ over $\mathbb{R}$. This is due to the fact that $\mathbb{C}$ can be viewed as a vector space of dimension 2 over $\mathbb{R}$ by interpreting $x + iy \in \mathbb{C}$ as $(x, y) \in \mathbb{R}^2$. The vector spaces $\mathbb{R}^{\mathbb{N}}$ of real sequences and $\mathbb{R}^{\mathbb{R}}$ of real real-valued functions are infinite-dimensional. Indeed, any finite subsequence of the vectors

$$(1, 0, \dots), (0, 1, 0, \dots), (0, 0, 1, 0, \dots), \dots$$

or of the functions

$$f_y : \mathbb{R} \to \mathbb{R}, x \mapsto \begin{cases} 1, & x = y, \\ 0, & x \neq y, \end{cases}$$

indexed by $y \in \mathbb{R}$, is comprised of linearly independent vectors (check this (Q3)).

A somewhat different, but equally interesting example of an infinite-dimensional vector space is $\mathbb{R}$ itself. In this case, we can take $\mathbb{Q}$ as the underlying field and consider $\mathbb{R}$ as a vector space over $\mathbb{Q}$, where addition is the addition of real numbers and scalar multiplication is multiplication by rational numbers. In this sense, $\mathbb{R}$ is an infinite-dimensional vector space. Indeed, if that were not the case, we would find a basis

1 Give some thought to how you would prove the uniqueness (Q2) of n drawing from your knowledge of elementary linear algebra.

$r_1, \ldots, r_n \in \mathbb{R}$ of some length n (the dimension) so that

$$\mathbb{R} = \left\{ \sum_{j=1}^{n} q_j r_j \,\middle|\, q_j \in \mathbb{Q} \text{ for } j = 1, \ldots, n \right\}.$$

This would yield a bijection between $\mathbb{Q}^n$ and $\mathbb{R}$, which cannot exist since $\mathbb{R}$ is not countable, whereas $\mathbb{Q}^n$ is. We included this example since it clearly points to the fact that $\mathbb{R}$ is an extremely large set, which "contains" a lot of structure. It is an excellent exercise to try and find an infinite number of linearly independent vectors in $\mathbb{R}$.[2]

3.2 Linear transformations

The structure of vector spaces is given by the availability of addition and scalar multiplication. This structure is typically called *linear* and vector spaces can be called linear spaces. This terminology is motivated by the fact that vectors, by fixing a direction, essentially determine a line: the vector $u \in V$ can be associated to the one-dimensional subspace $\mathbb{F}u$, which we imagine as a straight line. When considering maps $f : V \to W$ between two vector spaces V, W, it is possible to single out those, which preserve the linear structure. This means functions that map sums to sums and scalar multiples to scalar multiples. In mathematical words, this can be formulated as the validity of

$$f(\alpha u + \beta v) = \alpha f(u) + \beta f(v),$$

for any choice of $\alpha, \beta \in \mathbb{F}$ and $u, v \in V$. Notice that the addition and the scalar multiplications on the left of the identity are those of the space V, while they are those of the space W on the right-hand side. Maps between vector spaces satisfying this property are called *linear maps*. Clearly, not all maps are linear. A simple example is the map

$$f : \mathbb{R}^2 \to \mathbb{R}, \quad (x, y) \mapsto x^2 + y^2,$$

which does not satisfy the above linearity identity. The bijective map

$$C_{u_1,\ldots,u_n} : \mathbb{F}^n \to V, \quad \alpha = (\alpha_1, \ldots, \alpha_n) \mapsto \sum_{k=1}^{n} \alpha_k u_k$$

for a given basis $u_1, \ldots, u_n$ of the n-dimensional vector space V is an example of a linear map. It shows that an n-dimensional vector space over $\mathbb{F}$ looks like or can be identified

---

2 Hint: Take $r_k = \log(p_k)$ for $k \in \mathbb{N}$, where p_k is the kth prime number. Here, you can use (or verify for yourself) the known facts that there are infinitely many prime numbers, that every integer has a unique factorization into a product of integer powers of prime numbers, and that the square root of a prime is irrational.

with $\mathbb{F}^n$ itself. There are linear maps on infinite-dimensional spaces as well. Take the space $\mathbb{C}_c^{\mathbb{N}}$ of convergent sequences in $\mathbb{C}$ and consider the map

$$\lim : \mathbb{C}_c^{\mathbb{N}} \to \mathbb{C}, \quad u = (u_n)_{n\in\mathbb{N}} \mapsto \lim_{n\to\infty} u_n.$$

It is linear and you are asked to prove it. Similarly, we can take the space $\mathbb{R}^{\mathbb{R}}$ of real real-valued functions, fix a point $x_0 \in \mathbb{R}$, and set

$$\delta_{x_0} : \mathbb{R}^{\mathbb{R}} \to \mathbb{R}, \quad f \mapsto f(x_0),$$

i. e., the operation of evaluating functions at the point x_0. Again you are asked to show (Q4) that this is a linear map. If you are familiar with differentiability, you can take the vector space of real-valued continuously differentiable functions defined on the interval $(0, 1)$,

$$C^1((0,1),\mathbb{R}) = \{f : (0,1) \to \mathbb{R} \mid f' \text{ exists and is continuous}\},$$

and the linear map

$$D : C^1((0,1),\mathbb{R}) \to C^0((0,1),\mathbb{R}), \quad f \mapsto f',$$

which consists in producing the derivative of its argument, a continuously differentiable function f, and mapping it to its derivative f', which belongs to the space of continuous real-valued functions $C^0((0,1),\mathbb{R}) = C((0,1),\mathbb{R})$ defined on the same interval.

Next, we take a closer look at linear maps between finite-dimensional vector spaces. We shall come back to the infinite-dimensional case in later chapters. Consider a linear map $\mathcal{L} : V \to W$ between two finite-dimensional vector spaces of dimensions m and n, respectively. By choosing bases $u_1, \dots, u_m$ and $v_1, \dots, v_n$ for V and W, respectively, we can write any vector $u \in V$ as a linear combination of the basis elements, $u = \sum_{k=1}^m \alpha^k u_k$, where $\alpha^k = \alpha^k(u)$ depends (linearly) on u and is the kth coefficient in the expansion for $k = 1, \dots, m$. By linearity of the map, we have that

$$\mathcal{L}(u) = \mathcal{L}\left(\sum_{k=1}^m \alpha^k u_k\right) = \sum_{k=1}^m \alpha^k \mathcal{L}(u_k).$$

Now, for each $k \in \{1, \dots, m\}$, $\mathcal{L}(u_k) \in W$ and, therefore, has its own expansion in the chosen basis of W,

$$\mathcal{L}(u_k) = \sum_{j=1}^n L_k^j v_j$$

for uniquely determined coefficients $L^j_k \in \mathbb{F}$, $j = 1, \ldots, n$. We conclude that

$$\mathcal{L}(u) = \sum_{k=1}^{m} \alpha^k \left(\sum_{j=1}^{n} L^j_k v_j \right) = \sum_{j=1}^{n} \left(\sum_{k=1}^{m} L^j_k \alpha^k \right) v_j.$$

This means that the coefficients β in the basis expansion of $\mathcal{L}(u)$ can be computed by operating on the coefficients α in the expansion of u. You will recognize that this happens by matrix multiplication, since the coefficients β^j of the image vector are

$$\beta^j = \sum_{k=1}^{m} L^j_k \alpha^k = (L\alpha)^j, \quad j = 1, \ldots, n$$

for the matrix $L = [L^j_k]^{j=1,\ldots n}_{k=1,\ldots,m} \in \mathbb{F}^{n \times m}$ and $\alpha \in \mathbb{F}^m$. In conclusion, we see that, choosing bases for V and W, it is possible to describe vectors as tuples of numbers in the field and linear maps between V and W by matrices. In the example above, L is the matrix representation of the linear map $\mathcal{L}$ in these bases. This can be nicely summarized in the commutative diagram below, where we decorated some of the arrows with the symbol $\sim$ to indicate that the corresponding map is bijective.

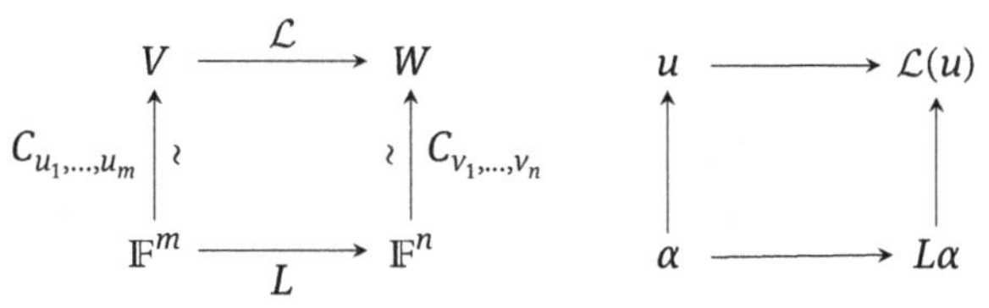

Denoting the vector space (check (Q5) that it is indeed one) of linear maps between V and W by $\mathcal{L}(V, W)$, the map

$$M : \mathcal{L}(V, W) \to \mathbb{F}^{n \times m}, \quad \mathcal{L} \to M(\mathcal{L}) = [L^j_m]^{j=1,\ldots n}_{k=1,\ldots,m}$$

is one-to-one and onto. Why did we choose to call the map M?

Notice that we are guilty of an abuse of notation here caused by the implicit choice of the basis consisting of the vectors

$$e_1 = \begin{bmatrix} 1 \\ 0 \\ \vdots \\ 0 \end{bmatrix}, \ldots, e_m = \begin{bmatrix} 0 \\ \vdots \\ 0 \\ 1 \end{bmatrix},$$

for the special vector space $\mathbb{F}^m$ ($m \in \mathbb{N}$). This basis is called the *natural basis* of $\mathbb{F}^m$. When considering a linear operator operator $\mathcal{L} : \mathbb{F}^m \to \mathbb{F}^n$ for $m, n \in \mathbb{N}$, we implicitly take the natural basis on the domain and on the range, thus conflating the linear map

$\mathcal{L}$ and its matrix representation L, which has columns $L_k^\bullet = \mathcal{L}(e_k) \in \mathbb{F}^n$ for $k = 1, \dots, m$. This, in turn, means that we think of any given matrix $L \in \mathbb{F}^{n\times m}$ as if it were the matrix representation of the linear map

$$L : \mathbb{R}^m \to \mathbb{R}^n, \quad \alpha \mapsto L\alpha,$$

obtained by matrix multiplication in the natural bases. An example can help clarify this point. The matrix

$$L = \begin{bmatrix} 1 & -1 \\ 1 & 1 \end{bmatrix}$$

is tacitly understood to be the matrix representation of

$$\mathcal{L} : \mathbb{R}^2 \to \mathbb{R}^2, \quad \alpha \mapsto L\alpha$$

with respect to the natural basis both on the domain and on the range. This means that

$$\mathcal{L}(e_1) = e_1 + e_2 \quad \text{and} \quad \mathcal{L}(e_2) = -e_1 + e_2.$$

However, this matrix could just as well be the representation matrix of a linear map $\overline{\mathcal{L}} : \mathbb{R}^2 \to \mathbb{R}^2$ in a basis other than the natural one (but still the same for both domain and range), say in the basis

$$v_1 = \begin{bmatrix} 1 \\ 0 \end{bmatrix}, \quad v_2 = \begin{bmatrix} 1 \\ 1 \end{bmatrix},$$

then it would encode the validity of

$$\overline{\mathcal{L}}(v_1) = v_1 + v_2 \quad \text{and} \quad \overline{\mathcal{L}}(v_2) = -v_1 + v_2,$$

and consequently that

$$\begin{aligned} \overline{\mathcal{L}}(e_1) &= \overline{\mathcal{L}}(v_1) = v_1 + v_2 = 2e_1 + e_2 \\ \overline{\mathcal{L}}(e_2) &= \overline{\mathcal{L}}(-v_1 + v_2) = -\overline{\mathcal{L}}(v_1) + \overline{\mathcal{L}}(v_2) = -2v_1 = -2e_1. \end{aligned}$$

This means that the matrix representation of $\overline{\mathcal{L}}$ in the natural basis (for both domain and range) would actually be given by

$$\begin{bmatrix} 2 & -2 \\ 1 & 0 \end{bmatrix}.$$

We see that the same matrix can represent different linear maps, depending on which underlying bases were chosen in its derivation. As a matter of fact, it is an interesting question to find bases in which the matrix representation of a given linear map is "simplest" or to determine which linear maps can be represented by the same matrix using different bases.

We conclude this section by looking at special linear maps that are related to the task of finding the simplest representation in the sense just explained. We will encounter them again in a later chapter. We keep V and W finite-dimensional but make one of them have dimension one. As we just saw, introducing a basis for V and W reveals that these vector spaces can be identified with $\mathbb{F}^m$ and $\mathbb{F}^n$ together with their natural bases, respectively, if their dimensions are m and n. We first assume that $m = 1$ and consider a matrix $L \in \mathbb{F}^{n\times 1}$ with a fixed but arbitrary $n \in \mathbb{N}$. Following the above discussion, L can be thought of as the linear map it defines by multiplication, and of which it is the matrix representation in the natural bases of $\mathbb{F}$ and $\mathbb{F}^n$. Since

$$L\alpha = \alpha L1 = \alpha v,$$

for $v = L1 \in \mathbb{F}^n$, the matrix L is thus fully determined by the vector v, which it "contains" as its sole column. In this way, we can identify $\mathbb{F}^{n\times 1}$ with $\mathbb{F}^n$.

Next, consider the case when $m \in \mathbb{N}$ is arbitrary and fixed, while $n = 1$. Again a matrix $L \in \mathbb{F}^{1\times m}$ (thought of as a linear map by multiplication) is completely determined by the vector β with components $\beta_k = Le_k = L^1_k$, where $k = 1, \dots, m$, laying flat in its only row. Indeed,

$$L\alpha = L\left(\sum_{k=1}^{m} \alpha^k e_k\right) = \sum_{k=1}^{m} \alpha^k Le_k = \sum_{k=1}^{m} \beta_k \alpha^k = \sum_{k=1}^{m} L^1_k \alpha^k.$$

This motivates the definition of the product between two vectors $\beta, \alpha \in \mathbb{R}^m$ given by

$$\beta^\top \alpha = \sum_{k=1}^{m} \beta_k \alpha^k,$$

where $\beta^\top$ denotes the "horizontal version" of the "vertical" vector β. We use subscripts (instead of superscripts) to identify the components of β in the product to reflect the "horizontal" use of β. This makes sense considering that β yields the sole row of the matrix $L \in \mathbb{F}^{1\times m}$ via

$$L = \beta^\top.$$

More in general, the *transpose* $M^\top \in \mathbb{F}^{m\times n}$ of a matrix $M \in \mathbb{F}^{n\times m}$ is defined as the matrix with entries

$$(M^\top)^k_j = M^j_k \quad \text{for } k = 1, \dots, m \text{ and } j = 1, \dots, n,$$

obtained by turning the columns (rows) of M into the rows (columns) of M^T. In this sense, we identify $\mathbb{F}^{1\times m}$ with $\mathbb{F}^m$(= "$(\mathbb{F}^m)^\top$") based on the identification of $\mathbb{F}^{m\times 1}$ with $\mathbb{F}^m$. Rows and columns can be viewed as building units of general matrices. Indeed, given a matrix $L \in \mathbb{F}^{n\times m}$ for general $m, n \in \mathbb{N}$, we can think of it either as consisting of n rows

$$\begin{bmatrix} L^1_\bullet \\ L^2_\bullet \\ \vdots \\ L^n_\bullet \end{bmatrix},$$

or of m columns

$$[L^\bullet_1 \quad L^\bullet_2 \quad \cdots \quad L^\bullet_m] .$$

In the first interpretation, we take a vector $\alpha \in \mathbb{F}^m$ and obtain the output vector $L\alpha$ collecting its components

$$(L\alpha)^j = L^j_\bullet \alpha = \sum_{k=1}^m L^j_k \alpha^k, \quad j = 1, \dots, n.$$

In the second, we use the components of α to build the linear combination of the columns given by $\sum_{k=1}^m \alpha^k L^\bullet_k = L\alpha$.

We can think of the simplest possible linear map between $\mathbb{F}^m$ and $\mathbb{F}^n$ as a map that takes the whole vector space $\mathbb{F}^m$, and maps it to a one-dimensional subspace of $\mathbb{F}^n$, necessarily of the form $\mathbb{F}v$ for some $v \in \mathbb{F}^n$. Since a vector $\alpha \in \mathbb{F}^m$ maps to a multiple of v, it is enough to know the appropriate stretching factor $\lambda = \lambda(\alpha) \in \mathbb{F}$. As we are considering a linear map, $\lambda : \mathbb{F}^m \to \mathbb{F}$ is linear (provide a proof) and is given by a matrix $\lambda \in \mathbb{F}^{1\times m}$ with a single row. In summary, such a simple linear map is fully determined by the two vectors $u \in \mathbb{F}^m$ and $v \in \mathbb{F}^n$, where $u = \lambda^\top$, the transpose of the row λ.

Using this notation, the simple linear map, call it L, has the representation

$$L\alpha = \lambda(\alpha)v = (u^\top \alpha)v = vu^\top \alpha, \quad \alpha \in \mathbb{F}^m,$$

i. e., $L = vu^\top$. Such a map is said to have *rank 1* or to be of *rank 1* to mean that it has a one-dimensional image space. Rank one maps can be thought of as basic building blocks of more general linear maps as follows from

$$L = \sum_{\substack{k=1,\dots,m \\ j=1,\dots,n}} L^j_k e^n_j (e^m_k)^\top, \quad L \in \mathbb{F}^{n\times m},$$

where $e_1^m, \dots, e_m^m$ and $e_1^n, \dots, e_n^n$ are the natural bases of $\mathbb{F}^m$ and $\mathbb{F}^n$, respectively. We will revisit this issue in Sections 8.1 and 8.2, where we will try and determine the "best" decomposition of a matrix into a sum of rank-one matrices.

3.3 The length of vectors

We saw in the previous chapter, when constructing real numbers, that one is often lead to consider or build sequences in order to solve a variety of problems and/or approximate their solutions. The verification of convergence will require the ability to measure distance, which is the basis for any concept of approximation (we need to know what "close" means). The mathematical concept of length for vectors is that of a *norm*. Given a vector space V, a norm on it is a map $|\cdot|_V : V \to \mathbb{R}$ satisfying:
(n1) $|u|_V \geqslant 0$ for $u \in V$ and equality only holds for the zero vector 0.
(n2) $|\alpha u|_V = |\alpha||u|_V$ for $\alpha \in \mathbb{F}$ and $u \in V$.
(n3) $|u + v|_V \leqslant |u|_V + |v|_V$ for $u, v \in V$.

Here, $|\alpha|$ denotes the absolute value (the modulus if $\mathbb{F} = \mathbb{C}$) of α in the field $\mathbb{F}$. Within the scope of this book, $\mathbb{F}$ will mostly be $\mathbb{R}$ or $\mathbb{C}$. As the notation indicates, a norm behaves similar to the absolute value in $\mathbb{R}$ or the modulus in $\mathbb{C}$. In particular, **(n3)** is the triangle inequality. We saw in Chapter 1 that assessing convergence requires a distance function. If a norm $|\cdot|_V$ is given on a vector space, then

$$d_V : V \times V \to \mathbb{R}, \quad (u, v) \mapsto d_V(u, v) = |u - v|_V$$

defines a metric, the metric induced by the norm $|\cdot|_V$. Remind yourself of what a metric is and verify this claim. Let $\mathbb{F} = \mathbb{Q}, \mathbb{R}$, or $\mathbb{C}$ and define $|\cdot|_p$ on $\mathbb{F}^m$ for $m \in \mathbb{N}$ and $p \in [1, \infty]$ by

$$|\alpha|_p = \begin{cases} \left(\sum_{k=1}^m |\alpha_k|^p\right)^{\frac{1}{p}}, & p \in [1, \infty), \\ \max_{k=1,\dots,m} |\alpha_k|, & p = \infty, \end{cases}$$

to obtain distinct norms (one for each p). The special norm obtained choosing $p = 2$ is called Euclidean, since when $m = 2$, it yields the length of Euclidean geometry based on Pythagoras' theorem. Indeed, if you interpret the vector $(1, 1)$ as the (directed) segment connecting the origin to the point $(1, 1)$ in the plane, Pythagoras's theorem yields a length of $\sqrt{1^2 + 1^2} = \sqrt{2}$ for it. While the above norms are all different, they do induce the same convergence for sequences in $\mathbb{F}^m$. This is due to the fact that, given any two of the norms corresponding to $p \neq q \in [1, \infty]$, it is possible to find a constant $C \geqslant 1$ such that

$$\frac{1}{C}|\alpha|_p \leqslant |\alpha|_q \leqslant C|\alpha|_p, \quad \alpha \in \mathbb{F}^m.$$

Verify (Q6) the validity of these inequalities as an exercise and convince yourself that the convergence of a sequence $(\alpha_n)_{n\in\mathbb{N}}$ in $\mathbb{F}^m$ to a limit $\alpha_\infty \in \mathbb{F}^m$ in any of these norms amounts to the convergence of each individual component sequence $(\alpha_n^k)_{n\in\mathbb{N}}$, where $k = 1, \dots, m$. In the process, try to make a connection between the above pair of inequalities and the shapes of the sets

$$\mathbb{B}_p(0,1) = \{x \in \mathbb{F}^n \mid |x|_p < 1\},$$

which are delimited by the sets $\{x \in \mathbb{F}^n \mid |x|_p = 1\}$ depicted in the figure below.

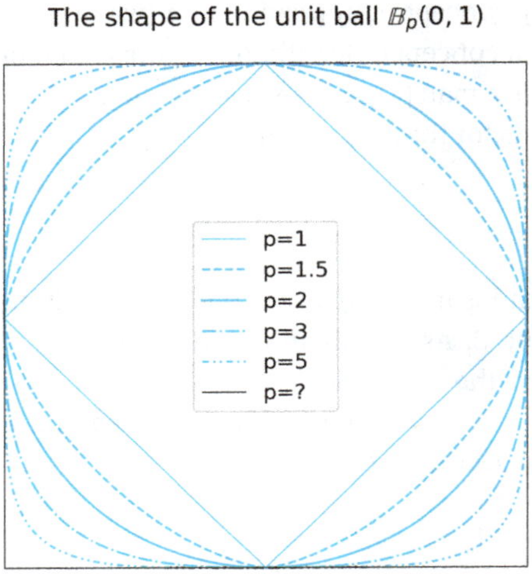

Things become more interesting when dealing with infinite-dimensional vector spaces. Consider first the space of sequences $\mathbb{R}^\mathbb{N}$. Extending the definition of the above norms by making the finite sum into an infinite one and setting

$$|\alpha|_p = \left(\sum_{n=1}^{\infty} |\alpha_n|^p\right)^{\frac{1}{p}}, \quad \alpha \in \mathbb{R}^\mathbb{N},$$

is not straightforward. Indeed, we would have $|\alpha|_p = \infty$ for the unbounded sequence $\alpha = (1, 2, 3, \dots)$ and any $p \in [1, \infty]$. Even the bounded sequence $(1, 1, 1, \dots)$ would have infinite norm for all $p \in [1, \infty)$. While it is possible to look for a norm (or at least a distance function) defined for all sequences, for most practical purposes, it is more convenient (and useful) to change the vector space. We can take the space of convergent sequences $\mathbb{R}_c^\mathbb{N}$ or that of Cauchy sequences $\mathbb{R}_{cs}^\mathbb{N}$, which is the same as $\mathbb{R}$ is complete, and use the $|\cdot|_\infty$ norm as a norm on them (check that the required properties are satisfied), since such sequences are necessarily bounded. By reducing the size of the vector space even further (when $p < \infty$), we can consider the vector spaces

$$\mathbb{R}_p^{\mathbb{N}} = \begin{cases} \{\alpha = (\alpha_n)_{n\in\mathbb{N}} \in \mathbb{R}^{\mathbb{N}} \mid \sum_{n=1}^{\infty} |\alpha_n|^p < \infty\}, & p \in [1,\infty), \\ \{\alpha = (\alpha_n)_{n\in\mathbb{N}} \in \mathbb{R}^{\mathbb{N}} \mid \sup_{n\in\mathbb{N}} |\alpha_n| < \infty\}, & p = \infty, \end{cases}$$

and use the corresponding $|\cdot|_p$ norm on them. Verification is left as an exercise, which might include researching the so-called Minkowsky inequality that amounts to the triangle inequality for these norms. Unlike the case of $\mathbb{R}^n$, these spaces are all different

$$\mathbb{R}_p^{\mathbb{N}} \neq \mathbb{R}_q^{\mathbb{N}} \quad \text{for } p \neq q,$$

and you are invited to investigate their relationships to one another (check which inclusions are valid). As an exercise also verify that $\mathbb{R}_c^{\mathbb{N}} = \mathbb{R}_{cs}^{\mathbb{N}}$ is a subspace of $\mathbb{R}_\infty^{\mathbb{N}}$ and that $\mathbb{R}_p^{\mathbb{N}}$ is one of $\mathbb{R}_c^{\mathbb{N}}$ for all $p \in [1,\infty)$. Can you show that the space $\mathbb{R}_p^{\mathbb{N}}$ is complete for $p \in [1,\infty]$? (Q7) If you are wondering why we are excluding $p \in (0,1)$, try to see why for yourself.[3]

Next, let us describe a distance function for the space $\mathbb{R}^{\mathbb{N}}$ of all sequences of reals

$$d(\alpha,\beta) = \sum_{n=1}^{\infty} \frac{|\alpha_n - \beta_n|}{1 + |\alpha_n - \beta_n|} 2^{-n}$$

Show that this is indeed a metric on $\mathbb{R}^{\mathbb{N}}$. What are the properties of the fractional term? Why do we use the factor 2^{-n}? Try (Q8) to understand and characterize the convergence that is induced by this metric. Convince yourself (Q9) that this distance does not come from a norm.

Moving on to examples of norms for vector spaces of functions such as $\mathbb{R}^{\mathbb{R}}$, we see that finding a norm cannot simply be handled by replacing a finite sum with an infinite one, since functions are vectors with uncountably many components. It is, however, possible to recycle one of the norms we already used, namely the so-called supremum norm. It amounts to

$$\|f\|_\infty = \sup_{x\in\mathbb{R}} |f(x)|,$$

and can be used on a variety of vector spaces such as, for instance,

$$\begin{aligned} \mathbb{R}_b^{\mathbb{R}} &= \{f : \mathbb{R} \to \mathbb{R} \mid f \text{ is bounded}\}, \\ \mathbb{R}_{bc}^{\mathbb{R}} &= \{f : \mathbb{R} \to \mathbb{R} \mid f \text{ is bounded and continuous}\}, \\ \mathbb{R}_{c,0}^{\mathbb{R}} &= \Big\{f : \mathbb{R} \to \mathbb{R} \;\Big|\; f \text{ is continuous and } \lim_{|x|\to\infty} f(x) = 0\Big\}. \end{aligned}$$

For all of these examples, verify that the supremum norm is indeed a norm. Infinite sums are reminiscent of integrals (the Riemann integral is defined as the limit of fi-

---

3 Hint: Which norm defining condition fails to hold?

nite sums, after all). This leads us to another class of examples. Let $a, b \in \mathbb{R}$ be given such that $a < b$ and consider the space of continuous all real- or complex-valued functions $f : [a, b] \to \mathbb{F}$, which we denote by $C([a, b], \mathbb{F})$ for short ($\mathbb{F} = \mathbb{R}, \mathbb{C}$). Continuous functions are Riemann integrable and, therefore, the norms

$$\|f\|_p = \left(\int_a^b |f(x)|^p \, dx \right)^{\frac{1}{p}}$$

are well-defined on $C([a, b], \mathbb{F})$ for any $p \in [1, \infty)$. This requires a proof (Q10), which you are encouraged to give. The validity of the triangle inequality, in this context, goes by the name of Minkowski integral inequality. Choosing $a = -1$ and $b = 1$, it can be verified that the sequence of functions $(f_n)_{n\in\mathbb{N}}$ given by

$$f_n(x) = \begin{cases} -1, & x < -\frac{1}{n}, \\ nx, & |x| \leqslant \frac{1}{n}, \\ 1, & x > \frac{1}{n}, \end{cases}$$

consists of continuous functions. The pointwise limit of this sequence is the function f_∞ with $f_\infty(0) = 0$, $f_\infty(x) = -1$ for $x < 0$, and $f_\infty(x) = 1$ for $x > 0$. This limiting function is not continuous and does therefore not belong to the space $C([a, b], \mathbb{R})$. It however holds that

$$\|f_n - f_\infty\|_p^p = 2 \int_0^{\frac{1}{n}} n^p x^p \, dx = \frac{2}{p+1} \frac{1}{n},$$

for $n \in \mathbb{N}$, which entails that

$$\|f_n - f_m\|_p \leqslant \|f_n - f_\infty\|_p + \|f_m - f_\infty\|_p \leqslant C\left(\frac{1}{n^{1/p}} + \frac{1}{m^{1/p}} \right), \quad m, n \in \mathbb{N}.$$

Since the right-hand side can be made arbitrarily small by taking m and n large enough for any $p \in [1, \infty)$, it follows that $(f_n)_{n\in\mathbb{N}}$ is a Cauchy sequence in the metric space $C([-1, 1], \mathbb{R})$ with distance (metric) defined by

$$d_p(f, g) = \|f - g\|_p = \left(\int_{-1}^1 |f(x) - g(x)|^p \, dx \right)^{\frac{1}{p}},$$

which is the metric induced by the $\|\cdot\|_p$ norm defined above. This shows that the metric space $(C([-1, 1], \mathbb{R}), d_p)$ is not complete in the sense of Chapter 1 since it admits Cauchy sequences, which do not converge to a limit in the space itself. Using the completion procedure of Chapter 1 via equivalence classes of Cauchy sequences, it is possible to obtain a complete space, which is called $L^p([-1,1])$. This is not the way this space is usually introduced but this approach is more elementary since it does not require measure theory. Of course, the counterexample to completeness can be adapted to any interval $[a,b]$ and completion leads to the space we denote by $L^p([a,b])$. Investigate the inclusion relations between the spaces $L^p([a,b])$ for different values of $p \in [1,\infty)$.

We conclude this section by mentioning an additional important and useful structure that is sometimes available on a vector space, that of scalar product. In the finite-dimensional case, when $\mathbb{F} = \mathbb{Q}, \mathbb{R}$, we set

$$(x|y) = x \cdot y = \sum_{k=1}^{n} x_k y_k \quad \text{for } x, y \in \mathbb{F}^n,$$

whereas we define

$$(z|w) = z \cdot w = \sum_{k=1}^{n} z_k \overline{w}_k \quad \text{for } z, w \in \mathbb{C}^n.$$

We recall the notation $\overline{z} = x - iy$ for the *complex conjugate* number of $z = x + iy \in \mathbb{C}$. Notice that $|z|^2 = z\overline{z}$ and that the complex definition of scalar product coincides with the real/rational one for vectors with vanishing complex parts and rational components. Among infinite-dimensional spaces, which admit a scalar product, we mention $\mathbb{F}_2^{\mathbb{N}}$ for $\mathbb{F} = \mathbb{R}, \mathbb{C}$ and $L^2([a,b])$, which were introduced above. Examples of scalar products on these spaces are

$$(x|y) = x \cdot y = \sum_{n=1}^{\infty} x_n \overline{y}_n \quad \text{for } x, y \in \mathbb{F}_2^{\mathbb{N}},$$

and

$$(f|g) = \int_a^b f(x)\overline{g(x)}\, dx \quad \text{for } f, g \in C([a,b], \mathbb{C}),$$

respectively. Notice that the latter scalar product can be extended to the completion $L^2([a,b])$ of the space of continuous functions since it generates the $\|\cdot\|_2$-norm in the sense that

$$\|f\|_2^2 = (f|f), \quad f \in L^2([a,b]).$$

3.3.1 Banach and Hilbert spaces

A vector space V with a norm $|\cdot|_V$ defined on it is called a *normed vector space*. It may or may not be complete. If it is, then it goes by the name of a *Banach space*. If it carries a scalar product, i. e., a map

$$(\cdot|\cdot)_V : V \times V \to \mathbb{F}$$

for $\mathbb{F} = \mathbb{R}, \mathbb{C}$ satisfying:
(sp1) $(v|u)_V = \overline{(u|v)_V}$ for $u, v \in V$,
(sp2) $(\alpha u + v|w)_V = \alpha(u|w)_V + (v|w)_V$ for $\alpha \in \mathbb{F}$ and $u, v, w \in V$,
(sp3) $(u|u)_V > 0$ for $u \neq 0$,

is called an *inner product space*. Given a scalar product $(\cdot|\cdot)_V$ on a vector space, we always obtain a norm on it by defining

$$|u|_V = \sqrt{(u|u)_V} \quad \text{for } u \in V.$$

Check that $|\cdot|_V$ is indeed a norm on V. Just as in planar geometry, a scalar product makes it possible to talk about the angle $\theta \in [0, \pi)$ between any two nonzero vectors $u, v \in V$ through

$$\cos(\theta) = \frac{(u|v)_V}{|u|_V |v|_V}.$$

In an inner product space, the concept of orthogonality can be introduced. Vectors $u, v \in V$ are called *orthogonal* if $(u|v)_V = 0$. Show (Q11) that any number of pairwise orthogonal (nontrivial) vectors $u_1, \dots, u_n$ in an inner product space are automatically linearly independent. The so-called *Cauchy–Schwarz inequality*

$$\left|(u|v)_V\right| \leqslant |u|_V |v|_V \quad \forall u, v \in V,$$

holds in inner product spaces. To prove it, first notice its validity whenever one of the vectors is zero. For $u \neq 0 \neq v$, consider

$$0 \leqslant (u + zv|u + zv)_V = |u|_V^2 + \bar{z}(u|v)_V + z(v|u)_V + z\bar{z}\,|v|_V^2,$$

and evaluate at $z = -\frac{(u|v)_V}{|v|_V^2}$ to obtain

$$\begin{aligned} 0 &\leqslant |u|_V^2 - \frac{\overline{(u|v)_V}(u|v)_V}{|v|_V^2} - \frac{(u|v)_V(v|u)_V}{|v|_V^2} + \frac{|(u|v)_V|^2}{|v|_V^2} \\ &= |u|_V^2 - \frac{|(u|v)_V|^2}{|v|_V^2}, \end{aligned}$$

using **(sp1)**. The proof is then concluded by simple algebraic manipulations. The metric of an inner product space is that induced by the norm $|\cdot|_V$ corresponding to the inner product $(\cdot|\cdot)_V$. A complete inner product space is called a *Hilbert space*.

Among the examples considered above,

$$(\mathbb{F}^n, |\cdot|_p), \quad \text{for } n \in \mathbb{N},\ p \in [1,\infty], \text{ and } \mathbb{F} = \mathbb{Q}, \mathbb{R}, \mathbb{C},$$
$$(\mathbb{F}_c^{\mathbb{N}}, \|\cdot\|_\infty), \text{ and } (\mathbb{F}_p^{\mathbb{N}}, \|\cdot\|_p) \quad \text{for } p \in [1,\infty), \text{ for } \mathbb{F} = \mathbb{Q}, \mathbb{R}, \mathbb{C},$$

are all normed vector spaces and they are complete when $\mathbb{F} \neq \mathbb{Q}$. When $p = 2$, they are inner product spaces (with respect to the inner product defined earlier) and complete, i. e., Hilbert spaces, if $\mathbb{F} \neq \mathbb{Q}$. Analogously,

$$(\mathbb{R}_b^{\mathbb{R}}, \|\cdot\|_\infty) \quad \text{and} \quad (\mathbb{R}_{bc}^{\mathbb{R}}, \|\cdot\|_\infty)$$

are Banach spaces, whereas

$$(\mathrm{C}([a,b], \mathbb{R}), \|\cdot\|_p)$$

are normed vector spaces for $p \in [1,\infty)$ (an inner product space for $p = 2$), and

$$\mathrm{L}^p([a,b])$$

are Banach spaces for $p \in [1,\infty)$ and a Hilbert space when $p = 2$. What is the natural inner product on $\mathrm{L}^2([a,b])$?

3.3.2 Gram–Schmidt orthogonalization algorithm

A finite-dimensional vector space V always has a basis $u_1, \ldots, u_n$, where n is its dimension. If V has an inner product $(\cdot|\cdot)_V$, the basis vectors may or may not be orthogonal to each other. When they are not, there is a procedure that produces an orthonormal basis out of them. A basis $v_1, \ldots, v_n$ of an inner product space V is called *orthonormal* iff

$$(v_j|v_k)_V = \begin{cases} 0, & j \neq k, \\ 1, & j = k, \end{cases} \qquad j, k \in \{1, \ldots, n\}.$$

Starting with an arbitrary basis $u_1, \ldots, u_n$, we can normalize u_1 to have unit norm and define

$$v_1 = \frac{u_1}{(u_1|u_1)_V^{1/2}} = \frac{u_1}{|u_1|_V}.$$

Next, we can use u_2 and v_1 to obtain a new unit vector v_2, which is orthogonal to u_1 in the form $u_2 + \alpha v_1$ by insisting that

$$(v_1|u_2 + \alpha v_1)_V = 0,$$

which yields $\alpha = -(v_1|u_2)_V$. Then we have that

$$v_2 = \frac{u_2 - (u_2|v_1)v_1}{|u_2 - (u_2|v_1)|_V}$$

has unit norm and is orthogonal to v_1. Notice that, geometrically, v_2 is obtained by removing the orthogonal projection of u_2 onto v_1 (and then normalizing the length to unit).

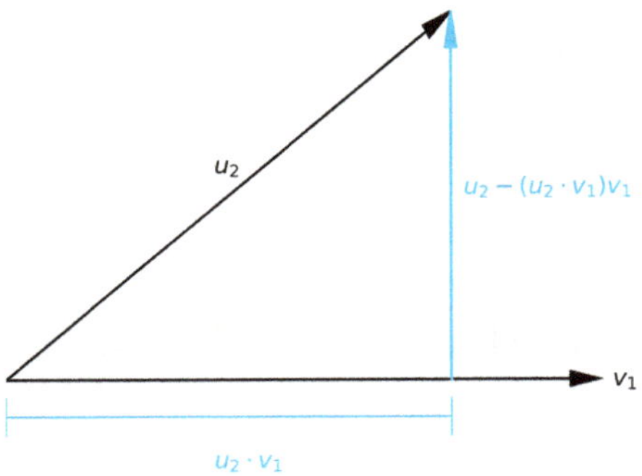

Assuming that we have already obtained orthonormal vectors $v_1, \ldots, v_k$ with $k < n$, we can similarly set

$$v_{k+1} = u_{k+1} + \sum_{j=1}^{k} \alpha_j v_j,$$

and insist that

$$\begin{aligned} 0 = (v_{k+1}|v_l)_V &= \left(u_{k+1} + \sum_{j=1}^{k} \alpha_j v_j | v_l \right)_V \\ &= (u_{k+1}|v_l)_V + \alpha_l \quad \text{for } l = 1, \ldots, k. \end{aligned}$$

This gives

$$v_{k+1} = \frac{u_{k+1} - \sum_{j=1}^{k}(u_{k+1}|v_j)v_j}{|u_{k+1} - \sum_{j=1}^{k}(u_{k+1}|v_j)v_j|_V}$$

and the procedure ends when $k + 1 = n$ and v_n is determined. The end result is an orthonormal basis given by $v_1, \ldots, v_n$.

3.4 Coordinates and coordinate functions

Given a finite-dimensional vector space V of dimension $d \in \mathbb{N}$ over $\mathbb{F}$, it possesses a basis $v_1, \dots, v_d$. This means that any vector $u \in V$ can be written as a unique linear combination

$$u = \sum_{k=1}^{d} x^k v_k,$$

for the appropriate $x \in \mathbb{F}^d$. This determines maps

$$X^k : V \to \mathbb{F}, \quad u \mapsto X^k(u) = x^k,$$

which we alternatively also denote by v^k or v_k' for $k = 1, \dots, d$. These are the coordinate maps, i. e., $v^k = v_k'$ yields the kth coordinate $v^k(u) = x^k$ of a vector in its expansion in the chosen basis. They are linear maps from the vector space V to the base field $\mathbb{F}$, i. e., elements of the vector space $V' = \mathcal{L}(V, \mathbb{F})$, also called *dual space* of V.

Fact. *The linear maps $v^1, \dots, v^d$ build a basis of V' whenever $v_1, \dots, v_d$ build one of V.*

In order to verify linear independence, assume that

$$\sum_{k=1}^{d} y_k v^k = 0 \quad \text{for } y_k \in \mathbb{F} \text{ and } k = 1, \dots d.$$

Thanks to the fact $v^k(v_j) = \begin{cases} 1, & j = k, \\ 0, & j \neq k \end{cases}$, the validity of which you are encouraged to check, this gives that

$$0 = \Big(\sum_{k=1}^{d} y_k v^k \Big)(v_j) = y_j,$$

for $j = 1, \dots, d$ and the stated linear independence follows. It remains to show that the vectors span the whole space. Let $v' \in V'$ and use linearity to see that

$$\begin{aligned} v'(u) = v'\Big(\sum_{k=1}^{d} x^k v_k \Big) &= \sum_{k=1}^{d} x^k v'(v_k) = \sum_{k=1}^{d} v^k(u) v'(v_k) \\ &= \Big(\sum_{k=1}^{d} v'(v_k) v^k \Big)(u), \end{aligned}$$

for any $V \ni u = \sum_{k=1}^{d} x^k v_k$ since $x^k = v^k(u)$. This shows that

$$v' = \sum_{k=1}^{d} y_k v^k$$

with $y_k = v'(v_k)$ for $k = 1, \dots, d$ and the argument is complete. $\surd$

When considering the Hilbert space $V = \mathbb{F}^d$ with an orthonormal basis $e_1, \dots, e_d$, you are asked to verify that $e^j = e_j'$ is given by $e_j^\top$. In general, using that $V = \mathbb{F}^d = \mathbb{F}^{d\times 1}$, it follows that $V' = \mathbb{F}^{1\times d}$, i. e., V' is a row space if we consider V a column space (and vice versa).

4 The topology of metric spaces

The main goal of this chapter is to introduce the important concept of compactness. Its importance stems from the fact that any sequence $(x_n)_{n\in\mathbb{N}}$ in a compact set always has at least a subsequence, which converges. By a subsequence, we mean a sequence that is obtained by (possibly) dropping some terms of the starting sequence. Compactness, or better, characterization of it, also represents a clear divide between finite- and infinite-dimensional spaces as we shall see in an example.

4.1 Open sets

The starting point is a metric space (M, d_M) and the concept of an open set. A set $O \subset M$ is called *open* if each of its points $x \in O$ admits a ball around it that is fully contained in O, i. e., if there is $r > 0$ (which depends on x) such that

$$\mathbb{B}_M(x, r) = \{y \in M \mid d_M(x, y) < r\} \subset O,$$

where $\mathbb{B}_M(x, r)$ is the so-called ball of radius r about the point x. Show that, for any $x \in M$ and $r > 0$, the ball $\mathbb{B}_M(x, r)$ is itself open (Q1). In $(\mathbb{R}, d_{|\cdot|}) = \mathbb{R}$, a simple example of an open set is given by any interval of the form

$$(a, b) = \{x \in \mathbb{R} \mid a < x < b\},$$

for $-\infty \leqslant a < b \leqslant \infty$. Verify that any metric space M itself is open, that finite intersections of open sets are open as are unions of any number of open sets. The limitation to finite intersections is necessary since the intersection of the intervals $(-1/n, 1+1/n) \subset \mathbb{R}$ for $n \in \mathbb{N}$ is the interval $[0, 1]$, a set that is not open since the points 0 and 1 do not admit any ball around them that is fully contained in $[0, 1]$. Sets which happen to be the complement of an open set are called *closed*. Intervals of the form

$$[a, b] = \{x \in \mathbb{R} \mid a \leqslant x \leqslant b\}$$

in $\mathbb{R}$ are examples of such sets. A set need not be open or closed as $[0, 1) \subset \mathbb{R}$ shows. The concept of open set is intimately connected to that of convergence that we did already define in Chapter 1. We leave it as an exercise to show that convergence of a sequence $(x_n)_{n\in\mathbb{N}}$ in M to a limit $x_\infty \in M$ is equivalent to the fact the ball $\mathbb{B}_M(x_\infty, r)$ of any radius $r > 0$ about x_∞ contains all but finitely many elements of the sequence. Given the definition of open set that we gave, this is the same as requiring that any open set containing x_∞ necessarily also contains all but finitely many terms of the sequence. Notice that the limit $x_\infty \in M$ of a convergent sequence $(x_n)_{n\in\mathbb{N}}$ in a closed subset C of a metric space necessarily belongs to C. Indeed, if $x_\infty \in C^c = M \setminus C$, which is open, then $r > 0$ can be found with $\mathbb{B}_M(x_\infty, r) \subset C^c$, and since this ball must contain all

https://doi.org/10.1515/9783110780925-004

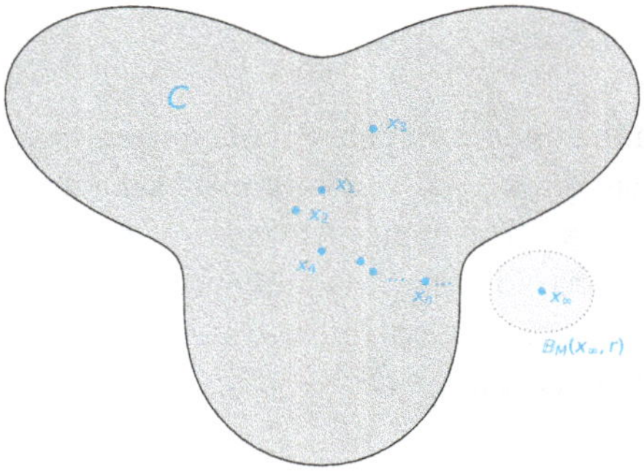

Figure 4.1: If the limit $x_\infty \in C^c$, then so is a ball $\mathbb{B}_M(x_\infty, r)$ of some radius $r > 0$. The latter would need to contain all but finitely many terms of the sequence, which are, however, all contained in the set C.

but finitely many of the terms of the sequence, we get a contradiction to the assumption that the sequence is in C. See Figure 4.1 for a depiction. You can think of this by saying that convergent sequences cannot escape a closed set. Convince yourself, by exhibiting an example (Q2), that the same is no longer true for nonclosed sets.

4.1.1 Open sets and continuity

A function $f : M \to N$ between metric spaces (M, d_M) and (N, d_N) is said to be *continuous at a point* $x \in M$ iff the following direct generalization of the same concept for real real-valued functions holds:

$$f(x) = \lim_{n\to\infty} f(x_n) \quad \text{for any } (x_n)_{n\in\mathbb{N}} \text{ in } M \text{ with } \lim_{n\to\infty} x_n = x.$$

Show that this is equivalent to the validity of

$$\forall\varepsilon > 0\ \exists\delta = \delta(x) > 0 \text{ s.t } d_N(f(x), f(y)) \leqslant \varepsilon \text{ provided } d_M(x, y) \leqslant \delta.$$

A function $f : M \to N$ that is continuous at all $x \in M$ is simply called *continuous*. If the continuity at $x \in U$ for a set $U \subset M$ does not depend on x, i. e., if δ can be chosen independently of $x \in U$, f is said to be *uniformly continuous* on U.

The use of open sets allows for a nice characterization of continuity. A function $f : M \to N$ is continuous if and only if $f^{-1}(O)$ is an open subset of M for any open subset O of N. To see this, we first fix an open set $O \subset N$ and set out to show that $f^{-1}(O) \subset M$ is open given that f is continuous. Take $x \in f^{-1}(O)$ so that $f(x) = y \in O$ and, by virtue of the fact that O is open, there is $\varepsilon > 0$ with $\mathbb{B}_N(y, \varepsilon) \subset O$. Then continuity implies the existence of $\delta > 0$ such that $f(\mathbb{B}_M(x, \delta)) \subset \mathbb{B}_N(y, \varepsilon)$, which is a rephrasing of the $\varepsilon - \delta$ condition above. Owing to $\mathbb{B}_N(y, \varepsilon) \subset O$, this shows that

$$\mathbb{B}_M(x,\delta) \subset f^{-1}(O).$$

Since $x \in f^{-1}(O)$ is arbitrary, this yields that $f^{-1}(O)$ is indeed open. The opposite direction is left as an exercise.[1]

4.2 Compactness and sequential compactness

We saw that a convergent sequence cannot escape a closed set. If convergence is not assumed and we take an arbitrary sequence $(x_n)_{n\in\mathbb{N}}$, there is no telling whether the sequence converges or whether even only a subsequence of it converges. The sequence $(n)_{n\in\mathbb{N}} = (1,2,3,\dots)$ contained in the closed set $\mathbb{R}$ (why is this set closed?) does not converge and it is also impossible to identify a subsequence that does. On the other hand, the sequence defined by

$$x_n = \begin{cases} 1, & \text{if } n \text{ is prime,} \\ 0, & \text{else} \end{cases}$$

does not converge but its subsequence $(x_n)_{\{n|n \text{ is prime}\}}$ does. We look for a general condition that ensures convergence of at least a subsequence that would work in a general metric space M. If a set $K \subset M$ has the property that any sequence in it always possesses a convergent subsequence, it is called *sequentially compact*. Owing to the characterization of convergence via open sets, it is enough to be able to show that there is a point $x_\infty \in M$ such that any ball of the form $\mathbb{B}_M(x_\infty, r)$ with arbitrary $r > 0$ contains infinitely many terms of the sequence. Of course, we can restrict our attention to sequences which, unlike the above example with the primes, do not have a term that is repeated infinitely many times, in which case it is trivial to find a convergent subsequence. Now, if a set $K \subset M$ had the property that whenever

$$K \subset \bigcup_{\lambda\in\Lambda} O_\lambda, \quad O_\lambda \text{ open for } \lambda \in \Lambda,$$

i. e., whenever K were covered by a family of open sets O_λ, there would be a finite subfamily that still covers K, i. e.,

$$K \subset \bigcup_{k=1}^{N} O_{\lambda_k},$$

for some $N \in \mathbb{N}$ and some $\lambda_1,\dots,\lambda_N \in \Lambda$, then a sequence $(x_n)_{n\in\mathbb{N}}$ in K would have infinitely many terms in at least one of the open sets O_{λ_k}, $k = 1,\dots,N$. If that were not the case, the sequence would not be infinite in the sense that it would not visit

1 Hint: Choose special open sets.

infinitely many distinct points. This motivates the definition of compactness for a set $K \subset M$, which reads: K is *compact* if any open cover always contains a finite subcover in the above sense.

Fact. *A subset $K \subset M$ of a metric space (M, d_M) is compact if and only if it is sequentially compact.*

We only prove that compactness implies sequential compactness since this is all we need for our purposes. Given a sequence $(x_n)_{n\in\mathbb{N}}$ in a compact set K, we therefore need to show that $x_\infty \in K$ can be found such that, for each $j \in \mathbb{N}$, there is $n_j \geqslant j$ with $x_{n_j} \in \mathbb{B}(x_\infty, \frac{1}{j})$. Without loss of generality, we can assume that the set $\{x_n \mid n \in \mathbb{N}\}$ is infinite, otherwise at least one of its elements would repeat infinitely many times in the sequence and thus would yield a convergent subsequence. We show that, assuming the contraposition, an open cover can be found, which has no finite subcover, yielding that K is not compact. Assume that, for each $y \in K$, we can find a natural number $j(y)$ with $x_n \notin \mathbb{B}(y, \frac{1}{j(y)})$ for $n \geqslant j(y)$. Since y runs through the whole set K, the family of open sets

$$\left\{\mathbb{B}\left(y, \frac{1}{j(y)}\right) \,\middle|\, y \in K\right\}$$

is an open cover of K, which cannot have a finite subcover. If it did, i. e., if

$$\{x_n \mid n \in \mathbb{N}\} \subset K \subset \bigcup_{k=1}^{m} \mathbb{B}\left(y_k, \frac{1}{j(y_k)}\right),$$

for some $y_1, \ldots, y_m \in K$ and some $m \in \mathbb{N}$, then at least one of these balls would necessarily contain infinitely many elements of the (infinite) set $\{x_n \mid n \in \mathbb{N}\}$, which is not possible by construction. √

We conclude this section with a fact about the interplay of continuity and compactness.

Fact. *A continuous function $f : M \to N$ between metric spaces maps compact sets to compact sets.*

In order to see this, take a compact $C \subset M$ and consider an open cover $\{O_\lambda \mid \lambda \in \Lambda\}$ of $f(C)$. By continuity, the sets $f^{-1}(O_\lambda)$ are open in M and build a cover of C. There must therefore be a finite subcover, i. e., it must hold that

$$C \subset \bigcup_{j=1}^{n} f^{-1}(O_{\lambda_j}),$$

for some $n \in \mathbb{N}$ and some $\lambda_j \in \Lambda$ for $j = 1, \ldots, n$. This implies that

$$f(C) \subset f\left(\bigcup_{j=1}^{n} f^{-1}(O_{\lambda_j})\right) = \bigcup_{j=1}^{n} O_{\lambda_j}$$

and yields a finite subcover for $f(C)$. This means that $f(C)$ is compact. √

As an exercise (Q3), show that a real-valued continuous function $f : M \to \mathbb{R}$ defined on a metric space, attains a maximum and a minimum on compact sets $C \subset M$. This means that, given a compact set $C \subset M$, there are $x_m, x_M \in C$ such that

$$\sup_{x\in C} f(x) = f(x_M) \quad \text{and} \quad \inf_{x\in C} f(x) = f(x_m).$$

4.3 Extension of uniformly continuous maps

In Chapter 2, we saw that $\mathbb{R}$ can be viewed as the completion of $\mathbb{Q}$ with respect to the absolute value. This means, in particular, that any real number can be approximated to an arbitrary degree of precision by a rational number. This relationship is described by saying that $\mathbb{Q}$ is a dense subset of $\mathbb{R}$. More in general, given a metric space (M, d_M), we say that a subset $D \subset M$ is *dense* in M if it holds that

$$\forall x \in M\ \forall \varepsilon > 0\ \exists z \in D \quad \text{with } d_M(x, z) \leqslant \varepsilon,$$

or more visually if $\mathbb{B}(x, \varepsilon) \cap D \neq \emptyset$ for every $x \in M$ and every $\varepsilon > 0$. It is a useful fact that uniformly continuous maps defined on a dense subset can be extended to a continuous map of the whole space provided the target space is complete.

Fact. *Let (M, d_M) and (N, d_N) be metric spaces. Assume that $D \subset M$ is dense in M and that N is complete. Then any uniformly continuous map $f : D \to N$ has a unique uniformly continuous extension $\bar{f} : M \to N$. We speak of an extension if $\bar{f}|_D = f$ holds.*

Letting $x \in M$, density of D yields a sequence $(x_n)_{n\in\mathbb{N}}$ in D satisfying $d_M(x, x_n) \to 0$ as $n \to \infty$ (why?). Given $\varepsilon > 0$, uniform continuity of f implies the existence of $\delta > 0$ such that

$$d_N(f(x_n), f(x_m)) \leqslant \varepsilon \quad \text{provided } d_M(x_n, x_m) \leqslant \delta.$$

Next, the latter inequality can be ensured by choosing $m, n \geqslant N$ for $N \in \mathbb{N}$ large enough since $(x_n)_{n\in\mathbb{N}}$ is a Cauchy sequence (as any convergent sequence is). Since $\varepsilon > 0$ is arbitrary in this argument, we see that $(f(x_n))_{n\in\mathbb{N}}$ is a Cauchy sequence in N. Thanks to completeness, a limit $y \in N$ exists, i. e.,

$$d_N(f(x_n), y) \to 0 \quad \text{as } n \to \infty.$$

We define $\bar{f}(x) = y$. For this definition to make sense, however, independence on the choice of the approximating sequence for x needs to be established. The details are left as an exercise (Q4). If $x \in D$, one can work with the constant sequence $(x, x, \dots)$ and set $\bar{f}(x) = f(x)$.

Finally, uniform continuity of $\bar{f}$ remains to be established. Let $\varepsilon > 0$ and take $\delta > 0$ so that

$$d_N(f(z), f(w)) \leqslant \varepsilon \quad \text{whenever } d_M(z, w) \leqslant 3\delta \text{ for } z, w \in D.$$

For any $x, y \in M$ given with $d_M(x, y) \leqslant \delta$, we can find $z, w \in D$ such that

$$d_M(z, x) \leqslant \delta, \quad d_M(w, y) \leqslant \delta$$

and, by the construction of $\bar{f}$, also such that

$$d_N(\bar{f}(x), f(z)) \leqslant \frac{\varepsilon}{3}, \quad d_N(\bar{f}(y), f(w)) \leqslant \frac{\varepsilon}{3}.$$

It follows that

$$d_M(z, w) \leqslant d_M(z, x) + d_M(x, y) + d_M(y, w) \leqslant 3\delta$$

and, therefore, that

$$d_N(\bar{f}(x), \bar{f}(y)) \leqslant d_N(\bar{f}(x), f(z)) + d_N(f(z), f(w)) + d_N(\bar{f}(y), f(w)) \leqslant \varepsilon.$$

Uniqueness is left as an exercise (Q5). $\surd$

We will use this fact in the next section to show the existence of linear functionals, which are not continuous.

4.4 Finite versus infinite-dimensional spaces

4.4.1 Compactness

The characterization of compactness marks a divide between finite and infinite-dimensional normed vector spaces. We recall that a normed vector space $(V, |\cdot|_V)$ always carries the structure of a metric space given by the induced distance d_V defined by $d_V(u, v) = |u - v|_V$ for $u, v \in V$. It therefore holds that compactness and sequential compactness are equivalent.

Fact (Bolzano–Weierstrass). *Let $(V, |\cdot|_V)$ be a finite-dimensional normed vector space over $\mathbb{R}$ or $\mathbb{C}$. A subset $K \subset V$ is compact if and only if it is closed and bounded.*

We give a proof for $V = (\mathbb{R}^d, |\cdot|_2)$, where $d \in \mathbb{N}$, and then for a general finite-dimensional vector space over $\mathbb{R}$. The complex case can be handled similarly. Given a closed and bounded subset $K \subset \mathbb{R}^d$, there is a constant such that

$$|x|_2 \leqslant M \quad \text{for all } x \in K.$$

Take a hypercube of the form $C_1^0 = [-R/2, R/2]^d$, which contains the closed ball $\overline{\mathbb{B}}_{|\cdot|_2}(0, M)$, and hence also K. Now consider a nontrivial sequence $(x_n)_{n\in\mathbb{N}}$ in K, i. e., sequence that visits infinitely many distinct points. Trivial sequences in this sense obviously have a convergent subsequence and the proof follows in that case. Split the above hypercube of side length R into 2^d subcubes C_j^1, $j = 1, \ldots, 2^d$ of side length $R/2$. One of these, say $C_{j_1}^1$, must therefore contain infinitely many terms of the sequence $(x_n)_{n\in\mathbb{N}}$. By splitting $C_{j_1}^1$ itself into 2^d subcubes $C_j^2 \subset C_{j_1}^1$, $j = 1, \ldots, 2^d$, of radius $R/2^2$, we similarly find a cube $C_{j_2}^2 \subset C_{j_1}^1$ containing infinitely many terms of the sequence. We can continue this construction indefinitely to obtain subcubes $C_{j_k}^k$ of radius $R/2^k$ containing infinitely many terms of the sequence. Finally, we choose $x_{n_k} \in C_{j_k}^k$ for each $k \in \mathbb{N}$ and obtain a Cauchy sequence since

$$|x_{n_k} - x_{n_l}|_2 \leqslant \sqrt{2^d}\frac{R}{2^{\min\{k,l\}}}, \qquad k, l \in \mathbb{N}.$$

Convince yourself that this is indeed a Cauchy sequence and explain why the factor $\sqrt{2^d}$ is needed in the inequality. Completeness of $\mathbb{R}^d$ implies that a limit $x_\infty \in \mathbb{R}^d$ exists for the subsequence $(x_{n_k})_{k\in\mathbb{N}}$ of $(x_n)_{n\in\mathbb{N}}$. Since K is closed, it must hold that $x_\infty \in K$. This yields sequential compactness, and hence compactness.

We still need to show that closure and boundedness are necessary for compactness. Indeed, if K is not closed, then there is a sequence in K with a limit outside of it. Such a sequence cannot possibly have a convergent subsequence with limit in K (why?). Similarly, if K is not bounded, then given any $k \in \mathbb{N}$, there is $x_k \in K$ with $|x_k|_2 \geqslant k$. Again it is not possible to find a convergent subsequence for such a sequence and again you are asked to explain why in more detail.

In order to deal with the case of a general real and finite-dimensional vector space V with norm $|\cdot|_V$, take a basis $u_1, \ldots u_n$ for it (see Section 3.1) and consider the bijective map

$$\Phi : \mathbb{R}^d \to V, \quad x \mapsto \sum_{j=1}^{d} x^j u_j.$$

This maps is continuous as follows from

$$|\Phi(x) - \Phi(y)|_V \leqslant \sum_{j=1}^{d} |x^j - y^j||u_j|_V \leqslant |x - y|_2 \left(\sum_{j=1}^{d} |u_j|_V^2 \right)^{1/2}$$

using the properties of a norm and the Cauchy–Schwarz inequality. Given a closed and bounded set $C \subset V$, the set $K = \Phi^{-1}(C)$ is therefore closed (why?). As for boundedness of K, assume that it is not. Then we can find a sequence $(x_n)_{n\in\mathbb{N}}$ in K such that $|x_n|_\infty \geqslant n$

for $n \in \mathbb{N}$. This implies that $(x_n^i)_{n\in\mathbb{N}}$ is not bounded for at least one component $1 \leqslant i \leqslant d$. It holds that

$$y_n^j = \frac{x_n^j}{|x_n|_\infty} \in [-1,1] \quad \text{for } j = 1,\dots,d \text{ and } n \in \mathbb{N}.$$

Notice that $y_n^j = \pm 1$ for at least one j (why?) and each $n \in \mathbb{N}$. It follows that there is at least one component $i \in \{1,\dots,d\}$ with either $y_n^i = -1$ or $y_n^i = 1$ for infinitely many indices n (why?). Using these indices, we obtain a convergent subsequence of $(y_n^i)_{n\in\mathbb{N}}$. Since $[-1,1] \subset \mathbb{R}$ is closed and bounded, we can extract further subsequences to ensure that all components converge along a common subsequence $(n_k)_{k\in\mathbb{N}}$, i. e., so that

$$y_{n_k}^j \to y^j \in [-1,1] \quad \text{for } l \to \infty \text{ and each } j = 1,\dots,d.$$

Since C is bounded, it holds that $|u|_V \leqslant M$ for $u \in C$ and some $M > 0$. Now, with $v_n = \Phi(x_n)$, we have that

$$0 \longleftarrow \frac{|v_{n_k}|_V}{|x_{n_k}|_\infty} = \left|\sum_{j=1}^d \frac{x_{n_k}^j}{|x_{n_k}|_\infty} u_j\right|_V \longrightarrow \left|\sum_{j=1}^d y^j u_j\right|_V,$$

and, since at least one of the coefficients y^j does not vanish, we also have that $|\sum_{j=1}^d y^j u_j|_V \neq 0$. The latter follows from the fact that the linear combination is not trivial and, therefore, cannot yield the zero vector (linear independence, remember?). This is a contradiction and boundedness follows. We conclude that K is compact and that so is $C = \Phi(K)$ as the image of the continuous map Φ of a compact set. Showing that a nonclosed or non-bounded subset of V cannot be compact can be done in a way completely analogous to the case when $V = \mathbb{R}^d$ was considered in the first part of the proof. The equivalence is established. $\surd$

While any compact set in a metric space will always be closed and bounded, these two properties are not sufficient for compactness in infinite-dimensional spaces. We refrain from proving this in general but give three examples. For one, take the Hilbert space $\mathbb{R}_2^{\mathbb{N}}$ and the vectors $e_k \in \mathbb{R}_2^{\mathbb{N}}$, $k \in \mathbb{N}$, with components $e_k^j = 0$ for $j \neq k$ and $e_k^k = 1$. These are orthogonal and have unit norm as follows from their definition and

$$(e_k|e_l) = \sum_{j=1}^\infty e_k^j e_l^j = \begin{cases} 0 & \text{if } l \neq k, \\ 1 & \text{if } l = k \end{cases}$$

so that

$$\|e_k - e_l\|_2 = \sqrt{2} \quad \text{for } l \neq k,$$

by Pythagoras' theorem. This shows that the sequence $(e_k)_{k\in\mathbb{N}}$ cannot have any convergent subsequence. In infinite-dimensional vector spaces, there are simply too many directions. It is an excellent exercise to think about what additional condition(s) could ensure compactness of sets in $\mathbb{R}_2^{\mathbb{N}}$. A similar approach can be taken for the space $\mathbb{R}_b^{\mathbb{R}}$ of bounded real real-valued functions with the supremum norm $\|\cdot\|_\infty$. Define the functions $i_{x_0} : \mathbb{R} \to \mathbb{R}$ by

$$i_{x_0}(x) = \begin{cases} 0, & x \neq x_0, \\ 1, & x = x_0, \end{cases}$$

and observe that $\|i_n - i_m\|_\infty = 1$ for $m, n \in \mathbb{N}$ with $m \neq n$. This clearly delivers a sequence $(i_n)_{n\in\mathbb{N}}$ in $\mathbb{R}_b^{\mathbb{R}}$, which does not have a convergent subsequence. This example is very similar to the one preceding it, in that it directly exploits the existence of infinitely many (linearly independent) directions in the space. We therefore give one additional example of a somewhat different nature in the space $\mathbb{R}_c^{[-1,1]}$ of continuous functions $f : [-1,1] \to \mathbb{R}$ with the supremum norm $\|\cdot\|_\infty$. Take the sequence $(f_n)_{n\in\mathbb{N}}$ defined by

$$f_n(x) = \begin{cases} -1, & x < -\frac{1}{n}, \\ nx, & x \in [-\frac{1}{n}, \frac{1}{n}], \\ 1, & x > \frac{1}{n}, \end{cases}$$

which converges pointwise to the discontinuous function f_∞ with

$$f_\infty(x) = \begin{cases} -1, & x < 0, \\ 0, & x = 0, \\ 1, & x > 0. \end{cases}$$

We show that no subsequence can converge with respect to the norm $\|\cdot\|_\infty$ defining the distance on $\mathbb{R}_c^{[-1,1]}$. If any subsequence were to converge, it would have f_∞ as a limit, but this is impossible since uniform limits (i. e., limits in the supremum norm $\|\cdot\|_\infty$) of sequence of continuous functions are necessarily continuous. In order to see the latter, take a point $x_0 \in [-1,1]$ and observe that

$$\begin{aligned} |f_\infty(x) - f_\infty(x_0)| &\leqslant |f_\infty(x) - f_n(x)| + |f_n(x) - f_n(x_0)| + |f_n(x_0) - f_\infty(x_0)| \\ &\leqslant \|f - f_n\|_\infty + |f_n(x) - f_n(x_0)| + \|f - f_n\|_\infty. \end{aligned}$$

Given convergence in the supremum norm, for any given $\varepsilon > 0$, it is possible to find $N = N(\varepsilon) \in \mathbb{N}$ such that

$$\|f_n - f_\infty\|_\infty \leqslant \frac{\varepsilon}{3} \quad \text{for } n \geqslant N,$$

and thus, in particular, for $n = N$. The continuity of f_N yields the existence of a positive $\delta = \delta(N) = \delta(N(\varepsilon))$ such that

$$|f_N(x) - f_N(x_0)| \leqslant \frac{\varepsilon}{3} \quad \text{for } x \in [-1,1] \text{ with } |x - x_0| \leqslant \delta,$$

which finally implies that

$$|f_\infty(x) - f_\infty(x_0)| \leqslant \frac{\varepsilon}{3} + \frac{\varepsilon}{3} + \frac{\varepsilon}{3} = \varepsilon,$$

provided $|x-x_0| \leqslant \delta$ and $x \in [-1,1]$. Since $x_0 \in [-1,1]$ was arbitrary, the same argument applies to any other point in the interval $[-1,1]$ and continuity of f_∞ follows. It is an excellent exercise (Q6) to (literally) challenge yourself to find conditions, which ensure sequential compactness in the space $\mathbb{R}_c^{[-1,1]}$. A good start consists in identifying what exactly is the breakdown in the above example. Counterexamples are always a good source of ideas in the search for conditions ensuring some mathematical property or other.

4.4.2 Linearity and continuity

Another fundamental and consequential difference between finite- and infinite-dimensional vector spaces manifests itself in the relation between linearity and continuity. Let us first look at linear maps between finite-dimensional spaces. Upon introduction of bases for the domain and the target spaces, we can assume without loss of generality that the linear map is defined on $\mathbb{R}^m$, has range $\mathbb{R}^n$, and is given by matrix multiplication with a matrix $L \in \mathbb{R}^{m\times n}$. While we have not proved it in this book, any two norms on a finite-dimensional space are equivalent.[2] Thus we can fix the Euclidean norm $|\cdot|_2$ on $\mathbb{R}^m$ and on $\mathbb{R}^n$. Then the Cauchy–Schwarz inequality gives that

$$|Lx|_2^2 = \sum_{i=1}^n \Big(\sum_{j=1}^m L_j^i x^j \Big)^2 \leqslant \sum_{i=1}^n \Big(\sum_{j=1}^m |L_j^i|^2 \Big) \Big(\sum_{j=1}^m |x^j|^2 \Big) = \|L\|_2^2 |x|_2^2,$$

if we set $\|L\|_2 = (\sum_{i=1}^n \sum_{j=1}^m |L_j^i|^2)^{\frac{1}{2}}$. This implies the estimate

$$|Lx - Ly|_2 \leqslant |L(x-y)|_2 \leqslant \|L\|_2 |x-y|_2, \quad x, y \in \mathbb{R}^m,$$

which amounts to continuity (in fact uniform continuity) of L.

Continuity of linear maps is no longer a given in infinite-dimensional spaces. We give an example to illustrate the nature of this phenomenon. Take the infinite-dimensional space defined by

2 Equivalence of two norms $|\cdot|_1$ and $|\cdot|_2$ on a vector space V amounts to the existence of a constant $c \geqslant 1$ with $\frac{1}{c}|x|_1 \leqslant |x|_2 \leqslant c\,|x|_1$ for every $x \in V$. Equivalent norms induce the same concept of convergence.

$$\mathbb{R}^{\mathbb{N}}_{00} = \{x \in \mathbb{R}^{\mathbb{N}} \mid x_n = 0 \text{ for } n \geqslant N \text{ for some } N \in \mathbb{N}\},$$

i. e., the space of sequences with at most finitely many non zero terms. It is infinite-dimensional since the vectors $\mathbb{R}^{\mathbb{N}}_{00} \ni e_j, j \in \mathbb{N}$, are linearly independent. The linear map

$$S : \mathbb{R}^{\mathbb{N}}_{00} \to \mathbb{R}, \quad x \to \sum_{n\in\mathbb{N}} x_n,$$

is well-defined and linear. Notice that the vector space $\mathbb{R}^{\mathbb{N}}_{00}$ can be given the norm defined by $|x|_\infty = \sup_{n\in\mathbb{N}} |x_n|$ and that it becomes a dense subspace of the Banach space

$$\mathbb{R}^{\mathbb{N}}_{0} = \left\{x \in \mathbb{R}^{\mathbb{N}} \mid \lim_{n\to\infty} x_n = 0\right\},$$

with respect to the same norm (prove this (Q7)). The linear map S is not continuous, since if it were, it would admit a unique uniformly continuous extension to $\mathbb{R}^{\mathbb{N}}_{0}$. This is, however, impossible since the sequence

$$x^m = (1, 1/2, \dots, 1/m, 0, 0, \dots), \quad m \in \mathbb{N},$$

in $\mathbb{R}^{\mathbb{N}}_{00}$ converges to the sequence $x^\infty = (\frac{1}{n})_{n\in\mathbb{N}} \in \mathbb{R}^{\mathbb{N}}_{0}$ as

$$|x^m - x^\infty|_\infty = \frac{1}{m+1} \to 0 \quad \text{as } m \to \infty,$$

but

$$S(x^m) = \sum_{n=1}^{m} \frac{1}{n} \to \infty \quad \text{as } m \to \infty.$$

4.5 Topology and convergence

In a metric space (M, d_M), convergence of a sequence $(x_n)_{n\in\mathbb{N}}$ to a limit $x_\infty \in M$ amounts to any open ball $\mathbb{B}_M(x_\infty, \varepsilon)$ (i. e., for any $\varepsilon > 0$) containing all but finitely many terms of the sequence. Thus, if the concept of open set is well-defined for subsets of a set X, one can attempt a definition of convergence by replacing "any open ball" by "any open set". In a metric space, open sets have the property that any union of any number of them still yields an open set as does any finite intersection, as well as that the whole space M and the empty sets are open. This motivates the following definition of the structure called *topology*. A collection τ_X of subsets of a given set X is called topology if and only if it satisfies the following conditions:

(t1) $X, \emptyset \in \tau_X$.

(t2) If $O_\lambda \in \tau_X$ for $\lambda \in \Lambda$ and any index set Λ, then $\bigcup_{\lambda\in\Lambda} O_\lambda \in \tau_X$.

(t3) If $O_j \in \tau_X$ for $j = 1, \ldots, n$ and $n \in \mathbb{N}$, then $\bigcap_{j=1}^n O_j \in \tau_X$.

Elements of the collection τ_X are called *open sets* and (X, τ_X) is called a *topological space*. In a topological space, it is possible to define convergence of a sequence $(x_n)_{n\in\mathbb{N}}$ to a limit $x_\infty \in X$ by requiring that each open set containing x_∞ also contain all but finitely many terms of the sequence. In a metric space, the collection of all sets that were defined to be open, i. e., of those sets for which each element admits a whole open ball (with positive radius) around them still fully contained in the set, do indeed build a topology. Given a set X, the largest topology one can think of is given by its power set 2^X. With respect to this topology, all sets are open. In particular, so are all singletons $\{x\}$ for $x \in X$. At the other end of the spectrum, we have the topology given by $\tau_X = \{\emptyset, X\}$. What concept of convergence is induced by these extremal topologies, or in other words, how do convergent sequences (Q8) look like with respect to these topologies?

Given a topological space (X, τ_X) and any subset Y of X, show (Q9) that the collection of sets

$$\tau_{X,Y} = \{O \cap Y \mid O \in \tau_X\} = \text{“}\tau_X \cap Y\text{”}$$

defines a topology on Y, the topology induced by τ_X on Y, and also called the relative topology. With this definition and considering $\mathbb{R}$ a metric space with respect to the distance induced by the absolute value, $d_\mathbb{R}(x, y) = |x-y|$ for $x, y \in \mathbb{R}$, convince yourself that $\tau_{\mathbb{R},\mathbb{N}} = 2^\mathbb{N}$. There are spaces, specifically vector spaces, on which it is possible to introduce a natural topology, which is not generated by a norm. By natural, we mean a topology, which captures an intuitive sense of convergence. Take for example, the space of real real-valued functions $\mathbb{R}^\mathbb{R}$ and define a set $O \subset \mathbb{R}^\mathbb{R}$ to be open if it is any union of sets of the form

$$\prod_{x\in\mathbb{R}} U_x$$

where $U_x \subset \mathbb{R}$ is open and $U_x = \mathbb{R}$ for all but finitely many $x \in \mathbb{R}$. Check that this is a topology. It yields pointwise convergence and it can be shown not to be induced by any norm. To see the former, take a sequence of functions $(f_n)_{n\in\mathbb{N}}$ in $\mathbb{R}^\mathbb{R}$. Convergence to $f_\infty \in \mathbb{R}^\mathbb{R}$ would amount to all but finitely many terms in the sequence to belong to any given open set of the above type. For this, it is enough that inclusion be valid for one of the sets in the union defining the specific open set chosen. This amounts to requiring that, for any given finite set $x_1, \ldots, x_M \in \mathbb{R}$ with $M \in \mathbb{N}$ and given $\varepsilon_1, \ldots, \varepsilon_M \in (0, \infty)$, there be $N \in \mathbb{N}$ with

$$|f_n(x_k) - f_\infty(x_k)| \leqslant \varepsilon_k \quad \text{for } k = 1, \ldots, M \text{ and } n \geqslant N.$$

This is always the case when the same holds true for $M = 1$ (why?), i. e., it is enough to consider single arguments $x \in \mathbb{R}$, which indeed amounts to pointwise convergence. Can you find a metric, which induces pointwise convergence on $\mathbb{R}^{\mathbb{R}}$?

While this is beyond the scope of this book, examples of topological spaces can be exhibited, the topology of which cannot even be induced by a metric. An example is the topological space just discussed.

Finally, notice that compactness can be defined in a topological space since it only requires the concept of open set. Clearly, different topologies on a given set determine different concepts of compactness, in general. It turns out that compactness is a stronger condition on a subset of a topological space than sequential compactness. Recall that the two concepts are equivalent in a metric space.

5 Differentiation

Differentiability is one of the central concepts of mathematics, and when available, makes it possible to simplify the analysis of many a problem. In this chapter, the classical concept of differentiability is introduced and carefully explained. The important concept of weak or generalized differentiability is (at least partially) discussed in Chapter 9. We will proceed in steps, starting with real real-valued functions, moving to maps between finite-dimensional vector spaces, infinite-dimensional spaces, and finally more general sets still. We strive to highlight two important points of view, one functional analytical and the other geometric (the distinction not being fully exclusionary). In both cases, a differentiable function is locally approximated by a linear map, but in the first the focus is squarely on the linear map, whereas in the second, it is on taking derivates along curves (i. e., in specific directions).

5.1 Introduction

The intuition behind differentiability is geometric: we think of a set as being differentiable at a point if it is locally well approximated by an affine space,[1] which is then necessarily tangent to the set at that point. In order to make this intuition mathematically sound, some care and effort are required. Let $f : \mathbb{R} \to \mathbb{R}$ be a given function and fix an argument $x_0 \in \mathbb{R}$. If f is not continuous at $x = x_0$, then it certainly cannot be well approximated by a line (i. e., by an affine function), which is continuous. Think of a function having a jump at x_0, for instance. Assuming continuity, we look for an affine function, i. e., for a function a of the form

$$a(x) = m(x - x_0) + b, \quad x \in \mathbb{R},$$

for $m, b \in \mathbb{R}$, that "well" approximates f near x_0. At a minimum, f and a should have the same value at x_0, which entails that

$$f(x_0) = l(x_0) = b \quad \text{and that} \quad a(x) = f(x_0) + m(x - x_0), \quad x \in \mathbb{R}.$$

We therefore would like to choose m so that

$$f(x) \simeq f(x_0) + m(x - x_0),$$

at least for $x \simeq x_0$. This suggests considering the expression

$$\frac{f(x) - f(x_0)}{x - x_0},$$

1 An affine space is a vector subspace translated inside the encompassing space. If $W \subset V$ for vector spaces V, W, then $v + W = \{v + w \mid w \in W\}$ is an affine space.

https://doi.org/10.1515/9783110780925-005

which is obtained solving for m. Unfortunately, this difference quotient depends on x and does not directly determine m in a unique way. Since we are only interested in a good approximation near x_0, we could try to take $x = x_0$. Unfortunately, we would obtain $\frac{0}{0}$, which is undefined. We know from basic calculus, however, that we can try and compute the limit

$$\lim_{x \to x_0} \frac{f(x) - f(x_0)}{x - x_0}$$

instead, which may or may not exist. It does not exist in general as the simple function $f(x) = |x|$ shows when $x_0 = 0$. If it exists, however, we can set

$$m = \lim_{x \to x_0} \frac{f(x) - f(x_0)}{x - x_0}$$

and obtain a good approximation for f at x_0, one the graph of which is in fact "tangent" to the graph of f. This motivates the definition of *differentiability* and of *derivative* for the function f at an argument x_0. The function f is called differentiable at x_0 if the limit exists, and in that case, its value is called derivative of f at x_0 and is denoted by $f'(x_0)$ or by $Df(x_0)$. The main reason for spelling out these details is to gain enough insight to be able to extend the definition of differentiability to more general useful contexts. Before continuing, we show that, for a differentiable function, the affine approximation just derived is the best possible one. Indeed, if a function $f : \mathbb{R} \to \mathbb{R}$ is differentiable at x_0 and $a(x) = f(x_0) + m(x - x_0)$ for $m \in \mathbb{R}$, then

$$\lim_{x \to x_0} \frac{f(x) - a(x)}{x - x_0} = \lim_{x \to x_0} \frac{f(x) - f(x_0)}{x - x_0} - m = Df(x_0) - m,$$

so that the approximation is best when $Df(x_0) - m = 0$ since

$$|f(x) - a(x)| \simeq |Df(x_0) - m||x - x_0|$$

for $x \simeq x_0$.

In order to simplify later discussions, it is convenient to introduce some notation. We say that $f(x) - f(x_0) = o(|x - x_0|^0) = o(1)$ as $x \to x_0$ to indicate the validity of

$$\lim_{x \to x_0} \frac{|f(x) - f(x_0)|}{|x - x_0|^0} = \lim_{x \to x_0} |f(x) - f(x_0)| = 0,$$

and say, analogously, that

$$f(x) - f(x_0) - Df(x_0)(x - x_0) = o(|x - x_0|^1) = o(|x - x_0|) \quad \text{as } x \to x_0,$$

to mean that

$$\lim_{x \to x_0} \frac{|f(x) - f(x_0) - Df(x_0)(x - x_0)|}{|x - x_0|} = 0.$$

In summary, differentiability amounts to "good" approximability by linear functions. In more intuitive words, differentiability at an argument x_0 means that the graph of the function would look like a straight line to you if you were a very tiny (infinitesimal) creature standing on it at the point $(x_0, f(x_0))$. We are tiny enough to perceive the earth as flat, which is the same experience a tiny germ might have on the surface of a soccer ball. On the other hand, if you are standing on the tip of pyramid (or a triangle, for that matter), it does not really matter how tiny you are, you will always be aware of two different slopes on either side. Thus, a nondifferentiable function will never look like a straight line, no matter how much you zoom in. Figure 5.1 visualizes this discussion.

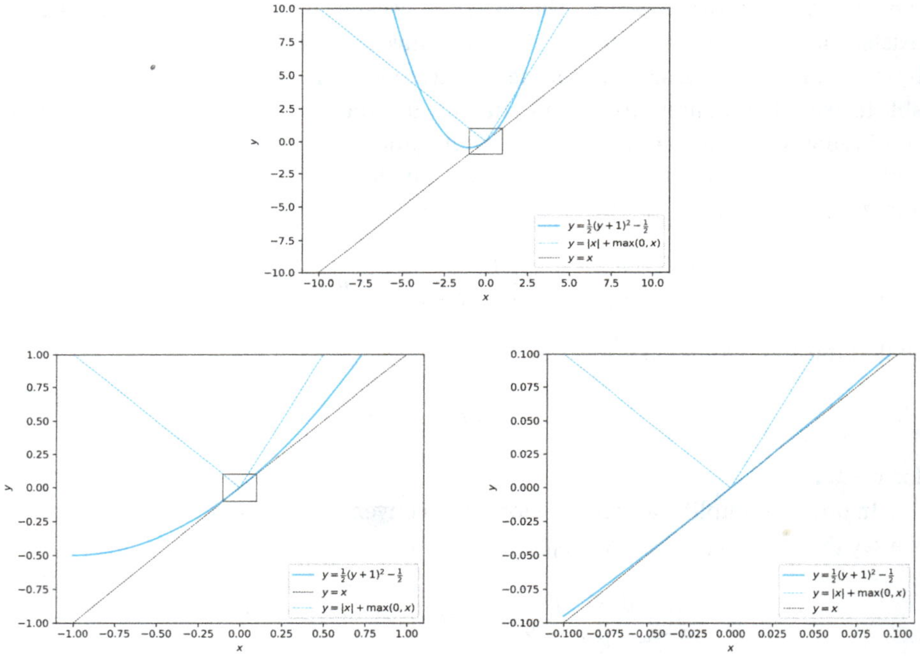

Figure 5.1: The graph of a differentiable function (blue line) is shown as well as that of a nondifferentiable function (dashed blue line). As one zooms in, the graph of the differentiable function appears to coincide more and more with that of its affine approximation (gray dashed line). This ostensibly does not occur for the nonsmooth function in that the kink always looks the same, regardless of magnification.

5.2 Derivatives as linear maps

Given normed vector spaces $(V, |\cdot|_V)$ and $(W, |\cdot|_W)$, which for now, we consider over the reals and finite-dimensional, we can think of linear maps as being particularly simple maps. In fact, by introducing bases $u_1, \dots, u_m$ in V and $v_1, \dots, v_n$ in W, we saw in Chapter 3, that any linear map $L : V \to W$ can be represented (fully understood) in terms of a matrix $M = M(L) \in \mathbb{R}^{n\times m}$, which maps, by matrix multiplication, the coefficients of arguments to the coefficients of their images. In the chosen bases, the matrix has entries given by

$$M^j_k = L(u_k)^j, \quad 1 \leqslant j \leqslant n,\ 1 \leqslant k \leqslant m,$$

where $L(u_k)^j \in \mathbb{R}$ denotes the jth coefficient in the basis expansion of the vector $L(u_k) \in W$. When $m = n = 1$ the matrix reduces to a real number and matrix multiplication to multiplication of reals. We take linear maps as the local model of differentiable maps as we did for real real-valued functions, meaning that a map $f : V \to W$ will be called differentiable at an argument $u_0 \in V$, if it can be well approximated by an affine map near u_0. By an affine map, we mean a function of the form

$$[u \mapsto L(u) + w],$$

where $L : V \to W$ is linear and $w \in W$ is a fixed vector in W. More precisely, f is considered differentiable at u_0 with derivative $Df(u_0) \in \mathcal{L}(V, W)$ if it holds that

$$\lim_{u\to u_0} \frac{|f(u) - f(u_0) - Df(u_0)(u - u_0)|_W}{|u - u_0|_V} = 0,$$

i. e., if

$$f(u) = f(u_0) + Df(u_0)(u - u_0) + o(|u - u_0|_V) \quad \text{as } u \to u_0,$$

borrowing from the notation introduced at the end of the previous subsection. Sometimes it is better to use the equivalent statement that

$$f(u_0 + h) = f(u_0) + Df(u_0)h + o(|h|_V) \quad \text{as } h \to 0,$$

where, throughout, we simplified the notation by writing Lu for $L(u)$ when $L \in \mathcal{L}(V, W)$ and $u \in V$.[2] Let us consider a couple of examples. Fix $A \in \mathbb{R}^{m\times m}$ and take the map

$$f : \mathbb{R}^m \to \mathbb{R}, \quad x \mapsto \frac{1}{2}x^\top Ax,$$

2 This is commonly done for linear maps on account that application of a linear map corresponds to multiplication by a matrix in the finite-dimensional context upon introduction of bases/coordinates.

and consider

$$\frac{1}{2}(x+h)^\top A(x+h) - \frac{1}{2}x^\top Ax - \frac{1}{2}x^\top(A+A^\top)h = \frac{1}{2}h^\top Ah,$$

which follows from linearity and using that $h^\top Ax = x^\top A^\top h$ for any $x, h \in \mathbb{R}^m$. The reader is reminded that the transpose $A^\top \in \mathbb{R}^{m\times m}$ of a matrix $A \in \mathbb{R}^{m\times m}$ has entries which satisfy $(A^\top)_j^k = A_k^j$ for $j = 1, \ldots, m$ and $k = 1, \ldots, m$. Next, observe that

$$|h^\top Ah| \leqslant |h|_2|Ah|_2 \leqslant |h|_2 \max_{i=1,\ldots,m} |A_\bullet^i|_2|h|_2 = C|h|_2^2,$$

by the Cauchy–Schwarz inequality. This implies that

$$0 \leqslant \frac{|f(x+h) - f(x) - \frac{1}{2}x^\top(A+A^\top)h|}{|h|_2} \leqslant C|h|_2 \longrightarrow 0 \quad (h \to 0),$$

showing that

$$f(x+h) - f(x) - \frac{1}{2}x^\top(A+A^\top)h = o(|h|_2)$$

and, therefore, that f is differentiable at x with derivative given by the linear map defined by

$$Df(x) : \mathbb{R}^m \to \mathbb{R}, h \mapsto \frac{1}{2}x^\top(A+A^\top)h.$$

For a symmetric matrix, i. e., for a matrix with $A^\top = A$, the derivative simplifies to $Df(x) = x^\top A$, which it generalizes the simple case when $m = 1$ and $A = a \in \mathbb{R} = \mathbb{R}^{1\times 1}$,

$$f(x) = \frac{1}{2}ax^2, \quad f'(x) = ax = x^\top a, \ x \in \mathbb{R},$$

since clearly $y^\top = y$ for any $y \in \mathbb{R}$.

The second example we consider is

$$f : \mathbb{R}^{n\times n} \to \mathbb{R}^{n\times n}, \quad M \mapsto M^2,$$

which consists in squaring a matrix. If we are to consider $\mathbb{R}^{n\times n}$ a normed vector space, we need to define a norm on it. A few options are available and we choose

$$\|M\| = \sup_{0\neq x\in\mathbb{R}^n} \frac{|Mx|_2}{|x|_2} = \sup_{|x|_2=1} |Mx|_2,$$

leaving it as an exercise (Q1) to verify the second identity, to show the validity of all norm defining properties, and to ponder what possible geometric interpretation this norm has. Now, if f is differentiable at a matrix M, the derivative $Df(M) : \mathbb{R}^{n\times n} \to \mathbb{R}^{n\times n}$ would have to be a linear map. Again notice that

$$(M+H)(M+H) - M^2 - MH - HM = H^2.$$

For matrices $M, N \in \mathbb{R}^{n\times n}$, it holds that

$$\|MN\| = \sup_{|x|_2=1} |MNx|_2 \leqslant \|M\| \sup_{|x|_2=1} |Nx|_2 \leqslant \|M\|\|N\|,$$

since $|My|_2 \leqslant \|M\||y|_2$ for any $y \in \mathbb{R}^n$. It follows that

$$\|f(M+H) - f(M) - (MH + HM)\| = \|H^2\| \leqslant \|H\|^2,$$

and, upon division by $\|H\|$, that

$$f(M+H) - f(M) - (MH+HM) = o(\|H\|) \quad \text{as } H \to 0.$$

We conclude that f is differentiable at any $M \in \mathbb{R}^{n\times n}$ and that the derivative at M is given by the linear map

$$Df(M) : \mathbb{R}^{n\times n} \to \mathbb{R}^{n\times n}, \quad H \mapsto MH + HM,$$

where linearity follows from the fact that matrix multiplication is linear (you should check this if you do not readily see it).

Finally, set $L^2 = L^2([0,1], \mathbb{R})$ and consider the map defined by

$$F : L^2 \to \mathbb{R}, \quad u \mapsto \frac{1}{2}\|u\|_2^2 = \frac{1}{2}\int_0^1 |u(x)|^2\,dx.$$

Observe that, for any given $h \in L^2$ it holds that

$$F(u+h) - F(u) = \int_0^1 u(x)h(x)\,dx + \frac{1}{2}\int_0^1 |h(x)|^2\,dx,$$

from which it follows that:

$$\frac{|F(u+h) - F(u) - \int_0^1 u(x)h(x)\,dx|}{\|h\|_2} = \frac{1}{2}\|h\|_2 \to 0 \quad \text{as } \|h\|_2 \to 0.$$

We conclude that F is differentiable at $u \in L^2$ with derivative there given by

$$Df(u) : L^2 \to \mathbb{R}, \quad h \mapsto \int_0^1 u(x)h(x)\,dx.$$

Notice that $F(u) = \frac{1}{2}(u|u)$ and that $DF(u) = (u|\cdot)$ in the sense that $DF(u)h = (u|h)$ for $h \in \mathrm{L}^2$.

5.3 Derivatives along curves

A real real-valued function can be thought of as a map of one variable and its derivative at an argument as the local rate of change of the function. In this sense, the function of one variable hardly needs to be defined along a straight line. As a matter of fact, x can be thought of as a parameter measuring the signed distance from an arbitrary origin along any line, straight or not. We define a *curvelet*, or simply a *curve*, in a normed vector space $(V, |\cdot|_V)$ over $\mathbb{R}$ as a differentiable map

$$\gamma : (-1, 1) \longrightarrow V,$$

and visualize it as the set $\{\gamma(t) \,|\, t \in (-1, 1)\}$ it traces in the space V and shown as a dashed line in the example image below.

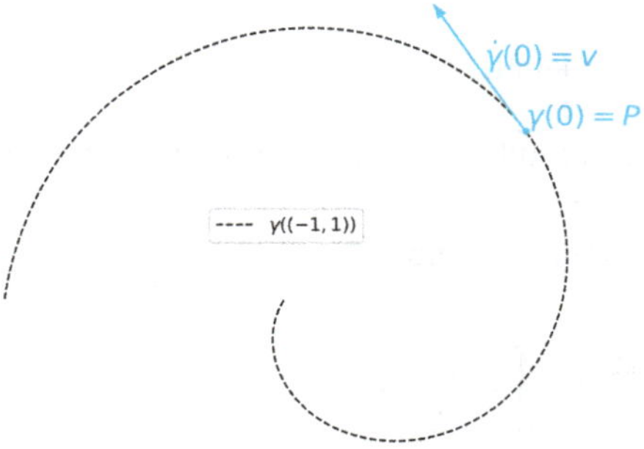

We saw in the previous section that we can think of the derivative $\dot{\gamma}(t) = D\gamma(t) \in V$ as the vector tangent to the curve in the point $\gamma(t)$ for any $t \in (-1, 1)$. In a normed vector space, vectors have a direction and a length, but do otherwise not have a "location". We can, however, think of the zero vector as a "point", in which case we can imagine any other vector $v \neq 0$ as originating in that special point and of its tip as the point determined by the vector v. If we think of a vector as the point at its tip in this sense, we use the different notation $P, Q, \ldots$. With this language, we can consider a curve γ in V that goes through a point P, i. e., satisfying $\gamma(0) = P$, and the tangent vector of which at the point P is given by $\dot{\gamma}(0) = v \in V$ (see depiction above). Given a function $f : V \to \mathbb{R}$, which happens to be differentiable at P, its local rate of change at the point P in direction v is then computed as

$$\frac{d}{dt}(f \circ \gamma)(0) = Df(\gamma(0))\dot{\gamma}(0) = Df(P)v,$$

by the chain rule. This is the *directional derivative* of f at P in direction v and is denoted by $\partial_v f(P)$. The use of directional derivatives reduces differentiability of functions of several variables to differentiability along specific directions, and thus to functions of a single variable. When $V = \mathbb{R}^m$, there are distinguished directions $e_1, \ldots, e_m$ determined by its natural basis. The so-called *partial derivatives* $\partial_{e_j} = \partial_j$ are therefore special directional derivatives. For a differentiable function $f : \mathbb{R}^m \to \mathbb{R}$, partial and directional derivatives at a point $x \in \mathbb{R}^m$ can be computed by

$$\partial_j f(x) = Df(x)e_j \quad \text{and by} \quad \partial_v f(x) = Df(x)v,$$

for $v \in \mathbb{R}^m$. In particular, partial and directional derivatives always exist for differentiable functions. The converse is, however, not true: the existence of all partial derivatives $\partial_j f(x), j = 1, \ldots, m$, of a function does not necessarily yield differentiability of f at x. It is shown in any basic advanced calculus class that equivalence can be regained if continuity of the partial derivatives is assumed. Here, we only present an example illustrating this phenomenon. Take the function $f : \mathbb{R}^2 \to \mathbb{R}$ defined by

$$f(x_1, x_2) = \begin{cases} \frac{x_1 x_2}{x_1^2 + x_2^2}, & \text{for } x \neq 0, \\ 0, & \text{for } x = 0, \end{cases}$$

and notice that $\partial_1 f(0) = \partial_2 f(0) = 0$ but f cannot be differentiable in $x = 0$ since it is not even continuous there as follows from

$$\lim_{t \to 0} f(t, 0) = 0 \neq \frac{1}{2} = \lim_{t \to 0} \frac{t^2}{2t^2} = \lim_{t \to 0} f(t, t).$$

5.4 Spaces of differentiable functions

We can now give more examples of infinite-dimensional vector spaces consisting of continuously differentiable functions. In order to efficiently deal with partial derivatives, we need to introduce so-called *multiindices*. These are simply vectors $\alpha = (\alpha_1, \ldots, \alpha_m) \in (\mathbb{N}_0)^m = \mathbb{N}_0^m$,[3] which are used to indicate the number of derivatives taken in the directions $e_1, \ldots, e_m$ for a function of m real variables. Specifically,

$$\partial^\alpha = \partial_1^{\alpha_1} \partial_2^{\alpha_2} \cdots \partial_m^{\alpha_m}.$$

The length of the multiindex $\alpha \in \mathbb{N}_0^m$ is denoted by $|\alpha|$ and is given by $\alpha_1 + \alpha_2 + \cdots + \alpha_m$. If $\Omega \subset \mathbb{R}^m$ is an open set and $m \in \mathbb{N}$, define

$$C(\Omega, \mathbb{R}^n) = \{f : \Omega \to \mathbb{R}^n \mid f \text{ is continuous}\}$$

---

3 The notation $\mathbb{N}_0$ stands for $\mathbb{N} \cup \{0\}$.

and, for $k \in \mathbb{N}$,

$$C^k(\Omega, \mathbb{R}^n) = \{f : \Omega \to \mathbb{R}^n \mid \partial^\alpha f \in C(\Omega, \mathbb{R}^n) \text{ for } |\alpha| \leq k\}.$$

While the spaces $C^k(\Omega, \mathbb{R}^n)$ are vector spaces (check), they do not carry a norm. Starting with $C((0,1), \mathbb{R})$, which contains the function $f(x) = \frac{1}{x}$, we see that functions in these spaces can be unbounded and prevent us from using natural candidates for a norm, such as the infinity norm given by

$$\|f\|_\infty = \sup_{x\in\Omega} |f(x)|.$$

If we add the condition of boundedness, we obtain normed vector spaces

$$BC^k(\Omega, \mathbb{R}^n) = \{f : \Omega \to \mathbb{R}^n \mid \partial^\alpha f \in BC(\Omega, \mathbb{R}^n) \text{ for } |\alpha| \leq k\},$$

starting from

$$BC(\Omega, \mathbb{R}^n) = \{f : \Omega \to \mathbb{R}^n \mid f \text{ is continuous and bounded}\}$$

if we use the infinity norm for the latter and

$$\|f\|_{k,\infty} = \sum_{|\alpha| \leq k} \|\partial^\alpha f\|_\infty$$

for the former. It is an excellent exercise (Q2) to show that the spaces $BC^k(\Omega, \mathbb{R}^n)$ are complete, i. e., Banach spaces.

5.5 The inverse and implicit function theorems

Given a function $f : \Omega \to \mathbb{R}^n$, defined on an open subset $\Omega \subset \mathbb{R}^n$, and $y \in \mathbb{R}^n$, we can ask whether $x \in \Omega$ can be found such that the equation $y = f(x)$ is satisfied. If the function were linear, i. e., of the form $f(x) = Ax$ for a matrix $A \in \mathbb{R}^{n\times n}$, this question would boil down to the invertibility of the matrix, which can be characterized by the nonvanishing of its determinant. Assume that we know that $f(x_0) = y_0$ for some specific argument $x_0 \in \Omega$ and we would like to determine whether a solution can still be found if y_0 is replaced by a close by value y. To simplify the discussion as much as possible, we can assume that $x_0 = 0$ be moving the origin of the coordinate system to that point and that $y_0 = 0$ be replacing the function f by the function $f - y_0$. Now, if the function happens to be differentiable, then we know that

$$\begin{aligned} f(x) &= f(x_0) + Df(x_0)(x - x_0) + o(|x - x_0|_2) \\ &= Df(0)x + o(|x|_2) \quad \text{as } x \to x_0 = 0. \end{aligned}$$

This suggests that, if the matrix $Df(0) \in \mathbb{R}^{n\times n}$ is invertible, then maybe the whole function is as well, at least for y and x near 0. This turns out to be true and goes by the name of *inverse function theorem*. The rest of this section is devoted to its derivation, which will rest on the use of a very useful fixed-point theorem named for the mathematician, Stefan Banach. Because of its independent interest, we formulate it in the more general context of a metric space and give it its own subsection.

5.5.1 Banach fixed-point theorem

Imagine that you are in Buenos Aires as a tourist, open up a city map on a café table while you are sipping on a maté, and ask yourself: is there a point on this map, which represents the very point in the city under it? The Banach fixed-point theorem gives a positive answer to this very question and also shows that there is only one such point.

Fact. *Let (M, d_M) be a complete nonempty metric space and let the map $f : M \to M$ be a contraction, i. e., a map satisfying*

$$d_M(f(x), f(y)) \leqslant \alpha d_M(x, y), \quad x, y \in M,$$

for a constant $\alpha \in (0, 1)$. Then f possesses a unique fixed point $x \in M$, i. e., a point such that $f(x) = x$.

It is a good exercise to translate between the example and the formulation of the theorem and determine what is playing the role of what. In particular, what do you think is the role of maté?[4]

In order to find a fixed point, choose any initial guess $x_0 \in M \neq \emptyset$ and recursively define a sequence $(x_n)_{n\in\mathbb{N}}$ in M by

$$x_{n+1} = f(x_n), \quad n \geqslant 0.$$

If we can show that it is a Cauchy sequence, completeness of M will imply that it converges to some limit $x_\infty \in M$, i. e., $d_M(x_n, x_\infty) \to 0$ as $n \to \infty$. Then the assumption yields

$$d_M(f(x_n), f(x_\infty)) \leqslant \alpha d_M(x_n, x_\infty) \to 0 \quad \text{as } n \to \infty,$$

showing that $x_{n+1} = f(x_n) \to f(x_\infty)$ as $n \to \infty$, and finally that necessarily $f(x_\infty) = x_\infty$. Toward proving the Cauchy property, observe that

$$d_M(x_n, x_{n-1}) \leqslant \alpha d_M(x_{n-1}, x_{n-2}) \leqslant \cdots \leqslant \alpha^{n-1} d_M(x_1, x_0) \quad \text{for } n \geqslant 2,$$

4 just joking.

from which follows that, for $n \geqslant m \geqslant 2$,

$$\begin{aligned}d_M(x_n, x_m) &\leqslant \sum_{j=m+1}^{n} d_M(x_j, x_{j-1}) \leqslant d_M(x_1, x_0) \sum_{j=m+1}^{n} \alpha^{j-1} \\ &= d_M(x_1, x_0)\left[\sum_{j=0}^{n-1} \alpha^j - \sum_{j=0}^{m-1} \alpha^j\right] \\ &= d_M(x_1, x_0)\left[\frac{1-\alpha^n}{1-\alpha} - \frac{1-\alpha^m}{1-\alpha}\right] = \frac{d_M(x_1, x_0)}{1-\alpha}(\alpha^m - \alpha^n).\end{aligned}$$

The claim now follows from the fact that $(\alpha^n)_{n\in\mathbb{N}}$ is Cauchy since it converges to 0 on account of $\alpha \in (0, 1)$. As for uniqueness, observe that, for two fixed-points x and $\tilde{x}$, we would have that

$$d_M(x, \tilde{x}) = d_M(f(x), f(\tilde{x})) \leqslant \alpha d(x, \tilde{x}),$$

and thus $0 \leqslant (1-\alpha)d_M(x, \tilde{x}) \leqslant 0$, from which it readily follows that $x = \tilde{x}$ since $1-\alpha \neq 0$.
√

5.5.2 The inverse function theorem

We derive the theorem first and formulate it at the end. Given an open set $\Omega \subset \mathbb{R}^n$ with $0 \in \Omega$ and $f : \Omega \to \mathbb{R}^n$ with $f(0) = 0$ satisfying some properties, which will be identified in the course of the derivation, we would like to show that the equation $f(x) = y$ is solvable for $y \simeq 0$ by a solution $x \simeq 0$. Notice that we cannot exclude the possibility that other solutions be found that are not close to the origin. To see the latter, simply consider $y = \sin(x)$, where $\sin(0) = 0$ and observe that, for each $|y| < 1$, there is a unique solution x with $|x| < \frac{\pi}{2}$, but infinitely many more farther from the origin. If we assume that f is differentiable at $x = 0$ and that $Df(0)$ is invertible, we can rewrite $y - f(x) = 0$ as

$$Df(0)x = Df(0)x + y - f(x)$$

or as the fixed-point equation

$$x = x + Df(0)^{-1}[y - f(x)] = \Phi_y(x).$$

It follows that, given y, any fixed point of Φ_y will deliver a solution of the equation $y = f(x)$. Since

$$\begin{aligned}|\Phi_y(x)|_2 &= |x + Df(0)^{-1}[y - f(x)]|_2 \\ &= |x + Df(0)^{-1}[y - Df(0)x + o(|x|_2)]|_2\end{aligned}$$

$$\leqslant \|Df(0)^{-1}\|[|y|_2 + o(|x|_2)] = c|y|_2 + o(|x|_2) \quad \text{as } x \to 0,$$

we find $\delta_1 > 0$, using the definition of o, such that

$$o(|x|_2) \leqslant \frac{|x|_2}{2} \leqslant \frac{\delta_1}{2} \quad \text{if } |x|_2 \leqslant \delta_1.$$

If $|y|_2 \leqslant \frac{\delta_1}{2c}$, then

$$|\Phi_y(x)|_2 \leqslant \frac{\delta_1}{2} + \frac{\delta_1}{2} = \delta_1 \quad \text{for } |x|_2 \leqslant \delta_1,$$

or in other words,

$$\Phi_y(\overline{\mathbb{B}}(0,\delta_1)) \subset \overline{\mathbb{B}}(0,\delta_1) \quad \forall y \in \mathbb{B}\left(0, \frac{\delta_1}{2c}\right).$$

If f is differentiable in a ball around $x = 0$, the mean value theorem[5] yields that

$$f(x_1) - f(x_2) = Df(x_2 + \tau(x_1 - x_2))(x_1 - x_2),$$

for some $\tau \in [0,1]$ and, if Df is continuous at $x = 0$, we have that

$$Df(x_2 + \tau(x_1 - x_2)) = Df(0) + o(1) \quad \text{as } x_1, x_2 \to 0.$$

Combining these, we obtain that

$$\begin{aligned} |\Phi_y(x_1) - \Phi_y(x_2)|_2 &= |x_1 - x_2 - Df(0)^{-1}(f(x_1) - f(x_2))|_2 \\ &\leqslant o(1)|x_1 - x_2|_2, \end{aligned}$$

as $x_1, x_2 \to 0$. Thus there is $\delta_2 > 0$ such that

$$|\Phi_y(x_1) - \Phi_y(x_2)|_2 \leqslant \frac{1}{2}|x_1 - x_2|_2$$

for $x_1, x_2 \in \overline{\mathbb{B}}(0,\delta_2)$ and any $y \in \mathbb{B}(0, \frac{\delta_1}{2c})$. This means that

$$\Phi_y : \overline{\mathbb{B}}(0,\delta) \to \overline{\mathbb{B}}(0,\delta)$$

is a contraction independently of $y \in \mathbb{B}(0, \frac{\delta}{2c})$, where $\delta = \min\{\delta_1, \delta_2\}$. Since $\overline{\mathbb{B}}(0,\delta)$ is complete as a closed subset of a complete space (why? (Q3)), we can apply the Banach fixed-point theorem to obtain

5 You can apply the mean value theorem to the function given by $[t \mapsto f(x_2 + t(x_1 - x_2))]$ and defined on the interval $[0,1]$.

$$\forall y \in \mathbb{B}\left(0, \frac{\delta}{2c}\right) \exists!\, x \in \overline{\mathbb{B}}(0,\delta) \quad \text{with } \Phi_y(x) = x \text{ or } f(x) = y.$$

Continuity of f implies that

$$\mathcal{U} = f^{-1}\left(\mathbb{B}\left(0, \frac{\delta}{2c}\right)\right)$$

is open, making $f|_{\mathcal{U}}$ injective and surjective on its range. Notice that unique solvability also yields a map $X = X(y)$, which satisfies $\Phi_y(X(y)) = X(y)$ for $y \in \mathbb{B}(0, \frac{\delta}{2c})$. Differentiating this identity with respect to y and denoting the identity matrix of size $n \times n$ by $\mathbb{1}_n$ gives

$$\begin{aligned} DX(y) &= D_y\Phi_y(X(y)) + D_x\Phi_y(X(y))DX(y) \\ &= Df(0)^{-1} + [\mathbb{1}_n - Df(0)^{-1}Df(X(y))]DX(y), \end{aligned}$$

which is equivalent to

$$Df(0)^{-1}Df(X(y))DX(y) = Df(0)^{-1},$$

and gives

$$DX(y) = Df(X(y))^{-1},$$

where invertibility of $Df(X(y))$ in a ball around 0 is ensured by the invertibility of $Df(0)$, the continuity of Df, and the continuity of the determinant, which implies that the set $\{M \in \mathbb{R}^{n\times n} \mid \det(M) \neq 0\}$ is open. This shows that $f|_{\mathcal{U}}^{-1}$ is differentiable and that

$$D(f|_{\mathcal{U}}^{-1})(y) = Df(x)^{-1} \quad \text{whenever } y = f(x).$$

We summarize our findings in the following.

Fact. *Let $\Omega \subset \mathbb{R}^n$ be open and $f : \Omega \to \mathbb{R}^n$ be differentiable in a ball around $0 \in \Omega$ with continuous derivative Df and satisfying $f(0) = 0$. If $Df(0)$ is invertible, then there is an open set $\mathcal{U} \subset \Omega$ containing 0 and an open set $\mathcal{V} \subset \mathbb{R}^n$ containing 0 such that*

$$f : \mathcal{U} \longrightarrow \mathcal{V},$$

is bijective, has a differentiable inverse, and

$$Df^{-1}(y) = Df(x)^{-1},$$

for $y = f(x) \in \mathcal{V}$.

It can be proved (try) that, if f is k-times continuously differentiable and satisfies the conditions of the inverse function theorem above, then so is $f^{-1} : \mathcal{V} \to \mathcal{U}$.

5.5.3 Newton's method revisited

We reconsider Newton's algorithm for finding zeros of a function $f : \mathbb{R}^n \to \mathbb{R}^n$ with new tools in our bag and show that it does indeed converge to a zero provided the initial guess is close enough to it. Recall that the algorithm consists in iterating via

$$x_{k+1} = x_k - Df(x_k)^{-1} f(x_k), \quad k \geqslant 0,$$

starting from an initial guess $x_0 \in \mathbb{R}^n$. We assume that f is twice continuously differentiable, denote the zero of f we are trying to compute by x_*, and observe that

$$\begin{aligned} x_{k+1} &= x_k - Df(x_k)^{-1}[f(x_*) + Df(x_*)(x_k - x_*) + O(|x_k - x_*|_2^2)] \\ &= x_* - Df(x_k)^{-1}[(Df(x_*) - Df(x_k))(x_k - x_*) + O(|x_k - x_*|_2^2)]. \end{aligned}$$

In order for this calculation to be valid and for the algorithm to be well-defined, it is necessary that $Df(x_k)$ be invertible. This can be ensured by assuming that $Df(x_*)$ be invertible and that $x_k \simeq x_*$ on account that Df is continuous. Since Df is once continuously differentiable by assumption, Df is Lipschitz continuous in any closed ball $\overline{\mathbb{B}}_2(x_*, R)$ about x_* (show this), so that

$$\|Df(x_*) - Df(x_k)\| \leqslant L|x_k - x_*|_2, \quad x_k \in \overline{\mathbb{B}}_2(x_*, R),$$

for some constant $L > 0$. Using the definition of O[6] and the boundedness of $(Df)^{-1}$ on $\overline{\mathbb{B}}_2(x_*, R)$, we can estimate the recursive formula above to obtain

$$|x_{k+1} - x_*|_2 \leqslant C|x_k - x_*|_2^2, \quad x \in \overline{\mathbb{B}}_2(x_*, R),$$

for a constant $C > 0$. Show that this last inequality is enough to conclude that the sequence $(x_k)_{k\in\mathbb{N}}$ converges to x_* for any initial guess x_0, provided x_0 is chosen close enough to x_*.

5.5.4 The implicit function theorem

Closely related to the inverse function theorem is the implicit function theorem to which it is equivalent.

Fact. *Consider a continuously differentiable function (i. e., differentiable with continuous derivative) $f : \Omega \to \mathbb{R}^n$, where $\Omega \subset \mathbb{R}^{n+m}$ is an open subset and $n, m \in \mathbb{N}$. Take a pair*

---

6 One says that $f(x) = O(g(x))$ as $x \to x_0$ for two functions f, g iff $|f(x)|_2 \leqslant C\,|g(x)|_2$ for $|x - x_0|_2 \leqslant \delta$ for some $\delta, C > 0$.

$(x_0, y_0) \in \mathbb{R}^n \times \mathbb{R}^m = \mathbb{R}^{n+m}$ with $f(x_0, y_0) = 0$ and assume that $D_x f(x_0, y_0)$ is invertible, where we set $D_x f = [\partial_{x_k} f^j]_{j,k=1,\dots,n}$. Then there are open sets $\mathcal{U} \ni (x_0, y_0)$ and $\mathcal{W} \ni y_0$ and a differentiable function $g : \mathcal{W} \to \mathbb{R}^n$ such that

$$f(g(y), y) = 0, \quad y \in \mathcal{W},$$

and this is the only solution of the form (x, y) in $\mathcal{U}$. Furthermore, it holds that

$$Dg(y) = -D_x f(g(y), y)^{-1} D_y f(g(y), y), \quad y \in \mathcal{W}.$$

Before giving a sketch of the proof, we consider a simple example as a first illustration. The real-valued function $f(x, y) = x^2 + y^2 - 1$ defined on $\mathbb{R}^2$ has derivative

$$Df(x, y) = [2x \quad 2y],$$

for which $D_x f(1, 0) = 2$ is invertible and $f(1, 0) = 0$. Thus there is an open set $\mathcal{W} \subset \mathbb{R}$ containing $y = 0$ and a map $g : \mathcal{W} \to \mathbb{R}$ such that $f(g(y), y) = 0$ for $y \in \mathcal{W}$. Clearly, in this case, the function is given by $g(y) = \sqrt{1 - y^2}$, and the sets by $\mathcal{W} = (-1, 1)$ and by $\mathcal{U} = \{(x, y) \mid x > 0\}$. It can also be verified that

$$Dg(0) = \frac{-2y}{2\sqrt{1 - y^2}}\bigg|_{y=0} = 0 = -\frac{D_y f(1, 0)}{D_x f(1, 0)}.$$

As for the proof, define the auxiliary function

$$F(x, y) = (f(x, y), y) : \mathbb{R}^{n+m} \to \mathbb{R}^{n+m},$$

which inherits the continuity and differentiability properties of f. It holds that $F(x_0, y_0) = (0, 0)$. By moving the origin of $\mathbb{R}^{n+m}$ to (x_0, y_0), this amounts to $F(0) = 0 \in \mathbb{R}^{n+m}$. Since

$$DF(x, y) = \begin{bmatrix} D_x f(x, y) & D_y f(x, y) \\ \mathbb{0} & \mathbb{1}_m \end{bmatrix}$$

the matrix $DF(0, 0)$ is invertible (since $D_x f(0, 0)$ is) and the inverse function theorem can be invoked and allows to conclude the proof. Filling in the missing details is an excellent exercise that requires a good understanding of the material.

5.6 A dive into coordinates

The use of coordinates is ubiquitous in mathematics, especially in differential geometry, and in applications. We illustrate their use and related issues in the "simpler" context of vector spaces. This sets the stage for their use in the description of differentiable manifolds while allowing for a more compact presentation, without sacrificing the main issues.

In an abstract finite-dimensional vector space V over $\mathbb{R}$, vectors cannot readily be thought of as tuples of numbers. Think of vectors in the physical world, such as representing a force acting on an object: while you can tell that a force is acting, you do not see tuples of numbers floating around. In order to be able to represent a vector by coordinates, it is necessary to choose a basis, i. e., to fix an appropriate number of linearly independent reference vectors.

Given a basis $u_1, \ldots, u_d$ of a $d \in \mathbb{N}$ dimensional real vector space V, we can use it to represent any vector $v \in V$ by its coordinates

$$X(v) = (X^1(v), X^2(v), \ldots, X^d(v)) \in \mathbb{R}^d$$

obtained from its unique representation as a linear combination of the chosen basis vectors

$$v = \sum_{j=1}^{d} X^j(v) u_j$$

With coordinates in hand, we can think of any curve γ in V as a curve in $\mathbb{R}^d$ given by its coordinate curve $X \circ \gamma$. Differentiability of the curve could then be defined as differentiability of $X \circ \gamma$, i. e., of all its coordinates

$$X^j \circ \gamma : (-1, 1) \to \mathbb{R}, \quad j = 1, \ldots, d.$$

Similarly, given a function $f : V \to \mathbb{R}$, it becomes possible to gain insight into its behavior by thinking of it as a function of the coordinates of its argument $v \in V$. We vary v by varying its coordinate vector x, which becomes the new independent variable. In other words, we consider f as the function

$$f \circ X^{-1} : \mathbb{R}^d \to \mathbb{R}, \quad x \mapsto f(X^{-1}(x)),$$

where it holds that

$$X^{-1} : \mathbb{R}^d \to V, \quad x \to \sum_{j=1}^{d} x^j u_j.$$

Differentiability of f could similarly be defined as differentiability of its coordinate representation $f \circ X^{-1}$. Of course, this raises a whole set of questions, including dependence on the arbitrary choice of basis performed in the beginning of this argument. We will return to this in a little bit. For now we point out that, using this idea of introducing coordinates, we effectively replaced the abstract vector space V with $\mathbb{R}^d$ so that curves in V and functions on V can be thought of as curves in and functions on $\mathbb{R}^d$. Given a curve γ in V and a function f on V, we can define differentiability directly as we did in the previous two sections provided we have a norm defined on V. We can, however, also take the coordinates' approach, which does not require a norm on V. In the latter case, if we are interested in the behavior of f along γ, we would consider $f \circ \gamma$ but would have to interpret this composition in coordinates as $(f \circ X^{-1}) \circ (X \circ \gamma)$ and compute its derivative as

$$\begin{aligned}\frac{d}{dt}&[(f \circ X^{-1}) \circ (X \circ \gamma)](t)\\ &= D(f \circ X^{-1})((X \circ \gamma)(t))\frac{d}{dt}(X \circ \gamma)(t)\\ &= \sum_{j=1}^{d}\left(\frac{\partial}{\partial x_j}(f \circ X^{-1})\right)((X \circ \gamma)(t))\frac{d}{dt}(X^j \circ \gamma)(t),\end{aligned}$$

which is perfectly justified since $f \circ X^{-1}$ and $X \circ \gamma$ are assumed to be differentiable. The outcome is the same as if we had computed

$$\left(\frac{d}{dt}f \circ \gamma\right)(t) = Df(\gamma(t))\dot{\gamma}(t),$$

since f and γ are differentiable. The reason of this seemingly vacuous exercise resides in the fact that the coordinates' approach can be used (as we shall see in Chapter 6) even in cases when it is impossible to define differentiability of f and γ directly due to a lack of a linear structure (and of a norm). The above discussion shows that the choice of particular coordinates does not affect the final result, i. e., that the choice of other coordinates (i. e., of another basis in this case) would have delivered the same outcome.

Working at a point P and using a curve γ in V with $\gamma(0) = P$ and $\dot{\gamma}(0) = v \in V$, it is possible to compute the directional derivative $d_P f(v)$ at P in direction v in coordinates X as

$$d_P f(v) = \sum_{j=1}^{d}\frac{\partial f}{\partial x^j}(X(P))dx^j(v) = \left(\sum_{j=1}^{d}\frac{\partial f}{\partial x^j}dx^j\right)(v),$$

where the two terms in the first sum are

$$\frac{\partial f}{\partial x^j}(X(P)) = \partial_j(f \circ X^{-1})((X \circ \gamma)(0))$$

and

$$dx^j(v) = \frac{d}{dt}(X^j \circ \gamma)(0) = X^j(\dot{\gamma}(0)) = X^j(v),$$

where the linearity of X was used in the second identity of the second term. As explained above, $d_P f(v)$ only depends on P, v, and f, i. e., it does not depend on the specific coordinates X used in the computation. This defines the so-called differential df of a function f, which encodes the local rate of change of the function f across any point P in any direction v. Of course, we have that $d_P f(v) = Df(P)v$.

We conclude this section by considering maps $f : V \to W$ between finite-dimensional real vector spaces. We obtain a coordinate representation of f by choosing bases $u_1, \ldots, u_m$ in V and $v_1, \ldots, v_n$ in W. Coordinates are given by maps $X : V \to \mathbb{R}^m$ and $Y : W \to \mathbb{R}^n$ yielding the uniquely determined coefficients in the respective basis expansions of vectors in these spaces. We can then think of f in coordinates as the map

$$Y \circ f \circ X^{-1} : \mathbb{R}^m \to \mathbb{R}^n, \quad x \mapsto Y(f(X^{-1}(x))),$$

which is often convenient to simply write as

$$(y^1, \ldots, y^n) = y = f(x) = (f^1(x^1, \ldots, x^m), \ldots, f^n(x^1, \ldots, x^m)),$$

by an abuse of notation. In this case, df can be thought of as

$$df = (d(Y^1 \circ f), \ldots, d(Y^n \circ f)) = (df^1, \ldots, df^n).$$

In this discussion, we considered functions between vector spaces, which can naturally be coordinatized by using globally defined linear maps (the maps X and Y above). As a matter of fact, coordinates are not even strictly necessary in this context since, at least when norms are available, differentiability can be understood directly as linear approximability as we explained in beginning of the chapter. The use of coordinates is, however, crucial when dealing with nonlinear spaces, such as manifolds introduced in the upcoming Chapter 6, where coordinates are, in general, only locally defined and not linear.

6 Manifolds

The basic idea behind the concept of a manifold is the use of $\mathbb{R}^n$ as a local model for more general sets, which do not carry a global linear structure like $\mathbb{R}^n$ does, but as in the case of differentiable manifolds, can locally be well approximated by a linear space. In this sense, a manifold can be thought of as a generalization of the graph of a differentiable function, which we know can be (well) approximated by the graph of an affine function. We will initially focus our attention on manifolds $M \subset \mathbb{R}^n$, i.e., manifolds that are subsets of $\mathbb{R}^n$, but the definition can be extended to the case where M is just a certain type of topological space to start with, which carries an additional differentiable structure based on maps (called charts) that coordinatize it in a very specific way. We will try to convey this point of view even as we mostly only consider the special case of submanifolds of $\mathbb{R}^n$ in the first part of this chapter.

6.1 A motivating example

Let us start by considering a unit circle as an example in order to motivate the general definition of a differentiable submanifold. We work in the plane $\mathbb{R}^2$ and consider the set $\mathbb{S}^1 = \{x \in \mathbb{R}^2 \mid |x|_2 = \sqrt{x_1^2 + x_2^2} = 1\}$. This set does not have a linear structure, but in the vicinity of any of its points, it is well approximated by a line, the one tangent to it at the point. It can, however, not be thought of as the graph of a single differentiable function, which prevents us from being able to introduce global coordinates to describe it. We can, however, do so locally. Let us start with coordinates: for the portion $\mathbb{S}^1_{x_2<0}$ of the circle where $x_2 < 0$, the map

$$X^1 : \mathbb{S}^1_{x_2<0} \to \mathbb{R}, \quad x \mapsto x_1,$$

defines local coordinates. In these coordinates, to each point P in $\mathbb{S}^1_{x_2<0}$, there corresponds exactly one point (the coordinate of P) in $(-1, 1) \subset \mathbb{R}$ and vice versa.

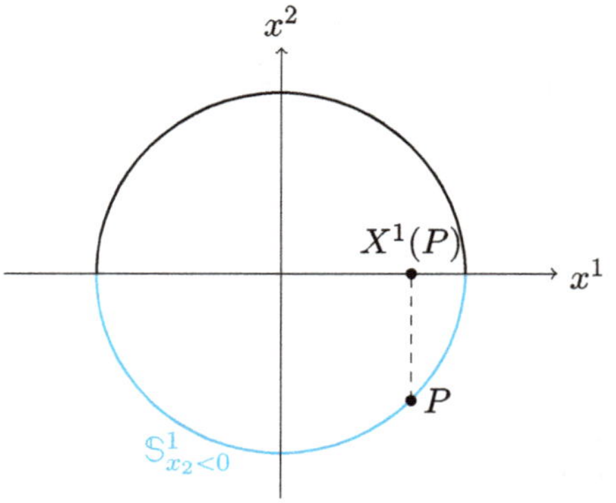

https://doi.org/10.1515/9783110780925-006

While the coordinate map X^1 can be defined on the whole set $\mathbb{S}^1$, taking any bigger open subset of $\mathbb{S}^1$ destroys the bijectivity of the map. The latter is a crucial ingredient if we are to avoid confusion in the way coordinates identify points in $\mathbb{S}^1$. We also see that $\mathbb{S}^1_{x_2<0}$ is the graph

$$\mathcal{G}_f = \{(x, f(x)) \mid x \in (-1,1)\}$$

of the differentiable function $f(x) = -\sqrt{1-x^2}$ for $x \in (-1,1)$. Because of this, we declare that $\mathbb{S}^1_{x_2<0}$ is differentiable. By using a similar construction, we can cover the whole set $\mathbb{S}^1$ by (open) subsets, each of which can be viewed as the graph of a differentiable function. For instance, we can take functions X^2, X^3, X^4 on $\mathbb{S}^1_{x_1>0}$, $\mathbb{S}^1_{x_2>0}$, and $\mathbb{S}^1_{x_1<0}$, respectively, with values given by $X^2(x) = X^4(x) = x_2$ and $X^3(x) = x_1$. These functions are all bijective. Another point of view is that of trying to generate the set $\mathbb{S}^1$ by a so-called parametrization, in this case as a curve. That is sometimes useful and yields a direct way to generate tangent vectors. In this case, we can use

$$\gamma : [0, 2\pi) \to \mathbb{R}^2, \quad \theta \mapsto (\cos(\theta), \sin(\theta)),$$

from which it is a direct calculation to obtain tangent vector

$$\dot{\gamma}(\theta) = (-\sin(\theta), \cos(\theta))$$

at the point $\gamma(\theta)$. Finally, we observe that $\mathbb{S}^1$ can also be viewed as the zero set of the smooth map $G : \mathbb{R}^2 \to \mathbb{R}$ given by $G(x_1, x_2) = x_1^2 + x_2^2 - 1$.

6.2 Submanifolds of $\mathbb{R}^n$

A differentiable submanifold M of $\mathbb{R}^n$ is first of all a subset. As such, it inherits the topology of $\mathbb{R}^n$, in the sense that

$$\tau_M = \{M \cap O \mid O \subset \mathbb{R}^n \text{ is open}\}$$

is a topology, and thus makes M into a topological space (check the validity of all axioms of topology). Now $M \subset \mathbb{R}^n$ is a *k-dimensional* C^1*-submanifold* if each of its points $x \in M$ admits an open set $O_x \in \tau_M$ containing x and a *homeomorphism* (i. e., a continuous map with continuous inverse)

$$\varphi_x : O_x \to \mathbb{B}_{\mathbb{R}^k}(0,1) = \{x \in \mathbb{R}^k \mid |x|_2 < 1\}$$

such that $\varphi_x^{-1} : \mathbb{B}_{\mathbb{R}^k}(0,1) \to \mathbb{R}^n$ is continuously differentiable and satisfies $\varphi_x(x) = 0$ and $\operatorname{rank}(D\varphi_x^{-1}(\cdot)) = k$, where the rank of a matrix is the maximal number of linearly independent columns, or equivalently, rows. The map φ_x provides local coordinates for M about the point x. While we insist that $\varphi_x(O_x) = \mathbb{B}_{\mathbb{R}^k}(0,1)$, the ball can be replaced by any open subsets of $\mathbb{R}^k$ without affecting the definition. We choose the ball for visualization purposes.

If

$$\mathbb{S}^{n-1} = \{x \in \mathbb{R}^n \mid |x|_2 = 1\},$$

then $\mathbb{S}^{n-1}$ is a $(n-1)$-dimensional C^1-submanifold of $\mathbb{R}^n$. Indeed, if $x \in \mathbb{S}^{n-1}$, define $\mathcal{P}_x = \{y \in \mathbb{R}^n \mid x^\top y = 0\}$, which is the hyperplane through the origin orthogonal to the vector x, and P_x to be the orthogonal projection onto $\mathcal{P}_x$. Taking an orthonormal basis $y_1, \ldots, y_{n-1}$ of $\mathcal{P}_x$ and denoting by $\eta = (\eta_1, \ldots, \eta_{n-1})$ the corresponding coordinates, define the (open) set $O_x = \{\tilde{x} \in \mathbb{S}^{n-1} \mid x^\top \tilde{x} > 0\}$ and

$$\varphi_x : O_x \to \mathbb{B}_{\mathbb{R}^{n-1}}(0,1), \tilde{x} \mapsto \eta(P_x(\tilde{x})).$$

We leave it to the reader to show that this map has the necessary properties (compute φ_x^{-1}). Since x is an arbitrary point of $\mathbb{S}^{n-1}$, this shows that we indeed have a submanifold.[1]

Fact. *A k-dimensional C^1-submanifold M of $\mathbb{R}^n$ can be viewed, in a small open region about any of its points, as the graph of a function*

$$f \in C^1(\mathbb{B}_{\mathbb{R}^k}(0,\delta), \mathbb{R}^{n-k})$$

for some $\delta > 0$.

Let $x \in M$ and let $\varphi_x : O_x \to \mathbb{B}_{\mathbb{R}^k}(0,1) \ni \xi$ be some local coordinates. The partial derivatives $\frac{\partial}{\partial \xi^j}\varphi^{-1}(0) = D\varphi^{-1}(0)_{\bullet j}$, $j = 1, \ldots, k$, yield linearly independent tangent vectors denoted by $u_1, \ldots, u_k$ to M at x, since

$$\xi_j \to \varphi^{-1}(0, \ldots, 0, \xi_j, 0, \ldots, 0), \quad (-1,1) \to \mathbb{R}^n$$

is a curve on M through x and $D\varphi^{-1}(0)$ has rank k. We complete $u_1, \ldots, u_k \in \mathbb{R}^n$ to a basis of $\mathbb{R}^n$ by adding $n-k$ vectors $u_{k+1}, \ldots, u_n$, which we can choose such that $u_i^\top u_l = 0$ for $i = 1, \ldots, k$ and $l = k+1, \ldots, n$ and such that $u_l^\top u_m = \delta_{lm}$[2] for $l, m = k+1, \ldots, n$. Using x as the origin and the vectors $u_1, \ldots, u_k$ as a basis for the tangent k-plane T_xM to M at x, gives coordinates $s' = (s^1, \ldots, s^k)$, i. e.,

1 Due to the rotational symmetry of the sphere, you can visualize the construction of φ_x by assuming, without loss of generality, that x is the north pole e_n.

2 We denote by δ_{lm} the so-called Kronecker symbol, which vanishes for $l \neq m$ and has value 1 for $m = l$.

$$T_xM = \left\{ x + \sum_{j=1}^{k} s^j u_j \;\middle|\; s^j \in \mathbb{R} \text{ for } j = 1, \ldots, k \right\}$$

and coordinates $s = (s^1, \ldots, s^n)$ for $\mathbb{R}^n$,

$$\mathbb{R}^n = \left\{ x + \sum_{j=1}^{n} s^j u_j \text{ for } j = 1, \ldots, n \right\},$$

as depicted below in the case $n = 2$, $k = 1$.

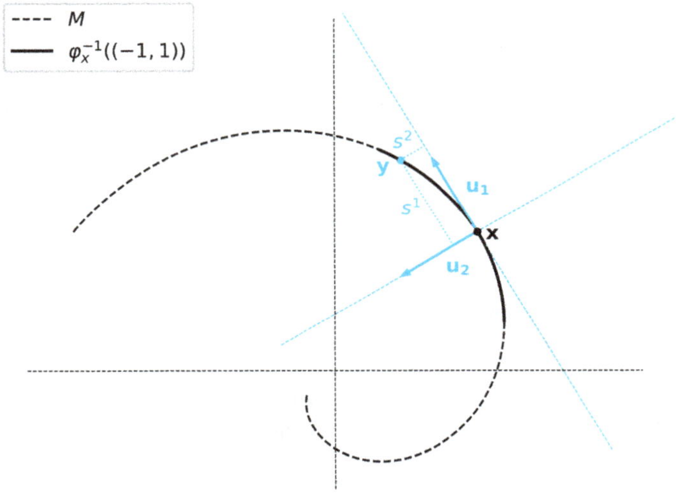

Denoting by $\pi_x : \mathbb{R}^n \to \mathbb{R}^k$, the linear map defined by $\pi_x(y) = s'(y)$ for $y \in \mathbb{R}^n$ with $y = x + \sum_{j=1}^n s^j u_j$, we consider the map

$$\pi_x \circ \varphi_x^{-1} : \mathbb{B}_{\mathbb{R}^k}(0,1) \to \mathbb{R}^k, \quad \xi \mapsto \pi_x(\varphi_x^{-1}(\xi)) = s'(\varphi_x^{-1}(\xi)),$$

for which we have that $\pi_x \circ \varphi^{-1}(0) = 0$ and $D(\pi_x \circ \varphi^{-1})(0) = \mathbb{1}_k$ (verify). The inverse function theorem can therefore be applied to obtain the invertibility of $\pi_x \circ \varphi_x^{-1}$ on the open set $(\pi_x \circ \varphi_x^{-1})^{-1}(\mathbb{B}_{\mathbb{R}^k}(0,\delta))$ for some $\delta > 0$. Denote the differentiable inverse of this map by ψ_x. Then, in s coordinates, we have that

$$\varphi_x^{-1}(\psi_x(\mathbb{B}_{\mathbb{R}^k}(0,\delta))) = \{(s', f(s')) \mid s' \in \mathbb{B}_{\mathbb{R}^k}(0,\delta)\},$$

where $f = s'' \circ \varphi_x^{-1} \circ \psi_x : \mathbb{B}_{\mathbb{R}^k}(0,\delta) \to \mathbb{R}^{n-k}$ and $s''(y)$ are the $(s^{k+1}, \ldots, s^n)$ coordinates of $y \in \mathbb{R}^n$. This shows that a small open region of M about x can be described as the graph of a differentiable function, which yields the claim since x was arbitrary. $\surd$

Being locally the graph of a differentiable function is in fact an equivalent definition of manifold. We just proved the more involved implication. How would you prove (Q1) the other?

A common source of submanifolds are zero sets of functions. We already saw the example of the unit sphere $\mathbb{S}^{n-1} \subset \mathbb{R}^n$, which can be thought of as the zero set of the function $f(x) = 1 - |x|_2^2$ defined for $x \in \mathbb{R}^n$. The dimension can vary, and before giving a general result, we consider another example. Define the submanifold of $\mathbb{R}^3$ as the unit circle $\mathbb{S}^1 \times \{0\}$ in the x_1x_2-plane. While it cannot be described by a scalar equation, it is possible to resort to a vector valued function such as

$$f : \mathbb{R}^3 \to \mathbb{R}^2, \quad x \mapsto (1 - x_1^2 - x_2^2, x_3),$$

for which $\mathbb{S}^1 \times \{0\} = f^{-1}(0)$. It is, however, not always the case that the zero set of a differentiable function is a manifold. Take $f : \mathbb{R}^2 \to \mathbb{R}$, $x \mapsto x_1x_2$, which has the union of the coordinate axis as its zero set

$$f^{-1}(0) = \mathbb{R} \times \{0\} \cup \{0\} \times \mathbb{R}.$$

This set is not a manifold in the above sense because it is impossible to describe it as the graph of a function in any region containing the origin. We give a sufficient condition for the zero set of a function to be, at least locally, a submanifold.

Fact. *Let $f \in C^1(\Omega, \mathbb{R}^{n-k})$ for an open set $\Omega \subset \mathbb{R}^n$ and assume that $f(x_0) = 0$ and that* $\operatorname{rank}(Df(x_0)) = n - k$. *Then the set $f^{-1}(0) \ni x_0$ is a k-dimensional submanifold of $\mathbb{R}^n$ in an open region about x_0.*

We can use the implicit function theorem as follows. Since the rank of $Df(x_0)$ is $n-k$, we can find $n-k$ of its columns, that are linearly independent. By reordering and relabeling the independent variables $x_1, \dots, x_n$, we can assume that it is the last $n-k$ columns (why exactly?). It follows that the matrix $D_{x''}f(x_0)$ is invertible if $x = (x', x'')$, where $x' = (x^1, \dots, x^k)$ and $x'' = (x^{k+1}, \dots, x^n)$. The implicit function theorem now yields open sets $\mathcal{U} \subset \mathbb{R}^n$, containing x_0, and $\mathcal{W} \subset \mathbb{R}^k$, containing x_0', such that any solution x of $f = 0$ in $\mathcal{U}$ has the form

$$x = (x', g(x')) \quad \text{for } x' \in \mathcal{W},$$

for a differentiable function $g : \mathcal{W} \to \mathbb{R}^{n-k}$. This shows that $f^{-1}(0)$ is indeed the graph of a differentiable function about x_0. √

The above (maximal) rank condition is sufficient, but not necessary as is demonstrated by considering the zero set of the function

$$f(x, y) = y^3 - x^5, \quad x, y \in \mathbb{R},$$

which is the graph $G = \{(x, \mathrm{sign}(x)|x|^{5/3}) \mid x \in \mathbb{R}\}$. Indeed, $Df(0,0) = [0 \quad 0]$ does not satisfy the rank condition but G is a 1-dimensional differentiable submanifold.

6.3 Abstract manifolds

Submanifolds of $\mathbb{R}^n$ are subsets that carry a topology, the one induced by the ambient space $\mathbb{R}^n$, and that are differentiable in the sense of being locally the graph of a differentiable function. An abstract manifold is also a topological space but it does not a priori lie inside any $\mathbb{R}^n$. The topological space (M, τ_M) is typically assumed to be Hausdorff. This means that two distinct points P and Q in M always possess neighborhoods[3] $\mathcal{U}_P$ and $\mathcal{U}_Q$, which are disjoint $\mathcal{U}_P \cap \mathcal{U}_Q = \emptyset$. We sometimes convey this by saying that points can be separated.

Fact. *A compact metric space is Hausdorff and any open set can be obtained as the countable union of open balls.*

Take $P \neq Q \in (M, d_M)$. Then $r = d_M(P, Q) > 0$ by one of the axioms of metric so that

$$\mathbb{B}_M(P, r/2) \cap \mathbb{B}_M(Q, r/2) = \emptyset,$$

for the open balls of radius $r/2$ about these points. As for the second property, notice that for each $n \in \mathbb{N}$ the collection of balls $\mathbb{B}_M(P, 1/n)$ for $P \in M$ yields an open cover of M. Due to its compactness, finite subcovers will exist: let $\{P_k^n \mid k = 1, \ldots, N_n\}$ be such that

$$M \subset \bigcup_{k=1}^{N_n} \mathbb{B}_M(P_k^n, 1/n).$$

Then

$$\mathcal{O} = \{\mathbb{B}_M(P_k^n, 1/m) \mid k = 1, \ldots, N_n, \text{ and } m, n \in \mathbb{N}\}$$

is a countable collection of open sets, which has the desired property. The reader is asked to show (Q2) than any open set in M can indeed be written as the union of a subcollection of $\mathcal{O}$. $\surd$

Now, even on an abstract manifold, we would like to be able to do calculus.[4] The basic ingredients will therefore be functions on M, which we would like to study and

---

3 A subset $\mathcal{U}$ of a topological space M is called neighborhood of a point P if it contains an open set containing P.

4 We do model our universe as a manifold, and since we are fully immersed in it, we can only perceive it as an abstract manifold and not as a submanifold of a linear space (what would that be anyway?).

curves, which we will need in order to take derivatives of functions. When dealing with a curve $\gamma : (-1,1) \to M$ on a submanifold $M \subset \mathbb{R}^n$, it is possible to consider its differentiability since it can be viewed as a curve $\gamma : (-1,1) \to \mathbb{R}^n$ in the vector space $\mathbb{R}^n$. Differentiable curves γ through a point $P \in M$ ($\gamma(0) = P$) can then be used to define tangent vectors $\dot{\gamma}(0)$ to M at P. This is not possible if M is just a topological space and it becomes a challenge to even define the concept of differentiable curve. As for functions $f : M \to \mathbb{R}$, we similarly need to define what it means for them to be differentiable. With this in mind, we attempt a definition of abstract manifold mimicking the one for submanifolds of $\mathbb{R}^n$. We could say that M is a k-dimensional C^1-manifold if each $P \in M$ has an open neighborhood $\mathcal{U}_P \subset M$, which admits a homeomorphism $\varphi_P : \mathcal{U}_P \to \mathbb{B}_{\mathbb{R}^k}(0,1)$ with $\varphi_P(P) = 0$. While this is well-defined, since it only involves continuity, yields local coordinates, and $\varphi_P^{-1} : \mathbb{B}_{\mathbb{R}^k}(0,1) \to M$ exists, differentiability cannot be directly enforced by requiring that φ_P^{-1} be continuously differentiable, as we did for submanifolds of $\mathbb{R}^n$, since M carries no linear structure.

We will now try to find a way to define a differentiable structure on M by focusing on the need to study the behavior of functions along curves. The idea will be quite natural if you think about navigation. In order to steer your ship to your destination, you use charts (maps) to trace your hopefully smooth (read differentiable) path (curve) through the ocean. Any single chart cannot cover the whole world, so we need a few of them covering different regions of the globe. Charts are usually collected into an atlas of the world.

Starting with a function f, say giving the distance to the next safe harbor, we try and trace a curve γ taking us to our destination, which always keeps us close enough to safety. To understand the curve and the function, we use charts, i. e., coordinates. If the interest is in the vicinity of the point P, we find ourselves at along our journey, we take a chart $X : \mathcal{U}_P \to \mathbb{B}_{\mathbb{R}^k}(0,1)$ defined in an open neighborhood of P and think of the function f in these coordinates as $f \circ X^{-1} : \mathbb{B}_{\mathbb{R}^k}(0,1) \to \mathbb{R}$ and call it differentiable when $f \circ X^{-1}$ is differentiable. Similarly, we call the curve γ differentiable if $X \circ \gamma$ is differentiable. This definitions manifestly depend on the choice of the coordinate X and we will need to address this issue. If differentiability holds in this sense, however, we can think of the behavior of f along γ, i. e., of $f \circ \gamma$ in coordinates, that is, through the study of $(f \circ X^{-1}) \circ (X \circ \gamma)$. Notice that the differentiability of $f \circ \gamma : (-1,1) \to \mathbb{R}$ can be considered since it is a function between vector spaces but the same cannot be said of $f : M \to \mathbb{R}$ nor $\gamma : (-1,1) \to M$. Thus coordinates represent a means of studying $f \circ \gamma$ as the composition of the function $f \circ X^{-1} : \mathbb{B}_{\mathbb{R}^k}(0,1) \to \mathbb{R}$ and the curve $X \circ \gamma : (-1,1) \to \mathbb{B}_{\mathbb{R}^k}(0,1)$ the differentiability of which can be discussed. If all we cared about was $f \circ \gamma$, we could content ourselves with the differentiability of $f \circ \gamma$. That would, however, allow for the possibility that, even for maps between vector spaces,

---

Having calculus available on an abstract manifold therefore helps us study the physics of the universe, for instance.

the composition be differentiable without both f and γ being so. Take, for instance, $f(x) = x^3$ and $\gamma(x) = \sqrt[3]{x}$, so that γ is not differentiable in $x = 0$ but $f \circ \gamma(x) = x$ is differentiable there. Also, this approach would avoid the use of charts, upsetting the captain of our ship, who only has access to charts.

Assuming the above differentiability, based on coordinates, we can then compute

$$\frac{d}{dt}(f \circ \gamma)(0) = \sum_{j=1}^{k} \frac{\partial}{\partial x^j}(f \circ X^{-1})(X \circ \gamma(0))\frac{d}{dt}(X^j \circ \gamma)(0).$$

If the differentiability of f and γ were possible to establish, then independence of the chosen coordinates would follow since the left-hand side could be evaluated by the chain rule to yield an expression that does not include any coordinates. Since that is not the case, independence needs to be addressed, otherwise the definition would not be very useful. To do so, take other coordinates $Y : \mathcal{V}_P \to \mathbb{B}_{\mathbb{R}^k}(0,1)$ about the point P with the aim of showing that

$$\sum_{j=1}^{k} \frac{\partial}{\partial x^j}(f \circ X^{-1})(X \circ \gamma(0))\frac{d}{dt}(X^j \circ \gamma)(0) = \sum_{j=1}^{k} \frac{\partial}{\partial y^j}(f \circ Y^{-1})(Y \circ \gamma(0))\frac{d}{dt}(Y^j \circ \gamma)(0).$$

Notice that

$$f \circ Y^{-1} = (f \circ X^{-1}) \circ (X \circ Y^{-1}),$$

and that

$$Y \circ \gamma = (Y \circ X^{-1}) \circ (X \circ \gamma).$$

Observe that the sets $Y(\mathcal{U}_P \cap \mathcal{V}_P)$ and $X(\mathcal{U}_P \cap \mathcal{V}_P)$ are open subsets of $\mathbb{B}_{\mathbb{R}^k}(0,1)$, since $\varphi_{Y,X}$ is a homeomorphism. Then, if the change of coordinates map

$$\varphi_{Y,X} = X \circ Y^{-1} : Y(\mathcal{U}_P \cap \mathcal{V}_P) \to X(\mathcal{U}_P \cap \mathcal{V}_P),$$

were differentiable, we would have that

$$\sum_{j=1}^{k} \frac{\partial f}{\partial y^j}\frac{d}{dt}(Y^j \circ \gamma) = \sum_{j=1}^{k}\sum_{l,m=1}^{k} \frac{\partial f}{\partial x^l}\frac{\partial X^l}{\partial y^j}\frac{\partial Y^j}{\partial x^m}\frac{d}{dt}(X^m \circ \gamma),$$

using the simplifying notation $\frac{\partial g}{\partial z^j}$ for $\frac{\partial}{\partial z^j}(g \circ Z^{-1})$ for local coordinates Z (X and Y in our case). Noticing that

$$\begin{aligned}\left[\sum_{j=1}^{k} \frac{\partial X^l}{\partial y^j}\frac{\partial Y^j}{\partial x^m}\right]_{l,m=1,\dots,k} &= D\varphi_{Y,X}D\varphi_{X,Y} = D\varphi_{Y,X}D\varphi_{Y,X}^{-1} \\ &= D\varphi_{Y,X}(D\varphi_{Y,X})^{-1} = \mathbb{1}_k,\end{aligned}$$

we obtain the validity of the desired identity and, with it, coordinate independence. For this to work, however, we need to assume that any change of coordinates map be differentiable. This understanding is now enough for making a viable definition of abstract manifold.

Toward the definition of k-dimensional C^m-manifold M for $m \in \mathbb{N}$, we let a C^m-*atlas* $\mathcal{A}$ on a Hausdorff and second countable topological space (M, τ_M) consist of a family $\{(\mathcal{U}_\lambda, X_\lambda) \mid \lambda \in \Lambda\}$ of *charts* (local coordinates) satisfying:

(m1) $X_\lambda(\mathcal{U}_\lambda) = \mathbb{B}_{\mathbb{R}^k}(0,1)$ for $\lambda \in \Lambda$.
(m2) $M \subset \bigcup_{\lambda\in\Lambda} \mathcal{U}_\lambda$.
(m3) $\varphi_{X_\lambda, X_\mu} \in C^m(X_\lambda(\mathcal{U}_\lambda \cap \mathcal{U}_\mu), X_\mu(\mathcal{U}_\lambda \cap \mathcal{U}_\mu))$ whenever $\mathcal{U}_\lambda \cap \mathcal{U}_\mu \neq \emptyset$ for $\lambda \neq \mu$.

Given an atlas $\mathcal{A}$ on M, a chart (U, X) is called compatible with it iff

$$\varphi_{X,X_\lambda} \in C^m(X(\mathcal{U} \cap \mathcal{U}_\lambda), X_\lambda(\mathcal{U} \cap \mathcal{U}_\lambda))$$

whenever $\mathcal{U} \cap \mathcal{U}_\lambda \neq \emptyset$. Two atlases are equivalent if all charts of one atlas are compatible with the other. This defines an equivalence relation of atlases on M and any equivalence class is called *differentiable structure* on M. A Hausdorff second countable topological space with k-dimensional C^m-differentiable structure is called *(abstract) k-dimensional C^m-manifold*. The dimension k of a manifold is often indicated along with the topological space as a superscript as in M^k. The simplest example of a k-dimensional C^m-manifold is $\mathbb{R}^k$ with the topology induced by $|\cdot|_2$ and the equivalence class of the atlas consisting of the single chart $X = (\mathbb{R}^k, \mathrm{id}_{\mathbb{R}^k})$.

Let now M^k be a C^m-manifold for $m \geqslant 1$ and $f : M \to \mathbb{R}$ be a real-valued function. It is called differentiable if $f \circ X^{-1} \in C^1(\mathbb{B}_{\mathbb{R}^k}(0,1), \mathbb{R})$ for all charts $(X, \mathcal{U})$ compatible with the atlas $\mathcal{A}$ defining its differentiable structure. Similarly, a curve $\gamma : (-1,1) \to M$ is called differentiable iff $X \circ \gamma$ is differentiable for all charts compatible with $\mathcal{A}$. Differentiable curves through a point $P \in M$ can be used to define the *tangent space $T_P M$* to M at P following the intuition that a differentiable curve through P has a direction (tangent vector) in which it passes through P. We introduce an equivalence relation for such curves through P via

$$\gamma_1 \sim \gamma_2 \Longleftrightarrow \frac{d}{dt}(f \circ \gamma_1)(0) = \frac{d}{dt}\bigg|_{t=0} f \circ \gamma_1 = \frac{d}{dt}\bigg|_{t=0} f \circ \gamma_2,$$

for all differentiable $f : M \to \mathbb{R}$. If $M = \mathbb{R}^n$, then this is equivalent to $\dot\gamma_1(0) = \dot\gamma_2(0)$ (check this (Q3)). Denote by $\mathcal{C}_P$ the set

$$\{\gamma : (-\delta, \delta) \to M \mid \delta > 0,\ \gamma \text{ is differentiable and } \gamma(0) = P\}$$

of differentiable curves through P and notice that it is convenient for curves to be defined in a variable (i. e., depending on the curve itself) open interval about the origin since we are only interested in the behavior (infinitesimally) close to P.

Fact. *The set of equivalence classes $C_P/\sim$ has the structure of a k-dimensional real vector space, it is denoted by T_PM and is called the tangent space to M at P.*

We need to define scalar multiplication and addition for equivalence classes $[\gamma]$ of curves, which satisfy the necessary properties. For $\alpha \in \mathbb{R}$, define

$$\alpha[\gamma] = [\gamma(\alpha \cdot)].$$

This is well-defined since $\gamma(\alpha \cdot) : (-\frac{\delta}{\alpha}, \frac{\delta}{\alpha}) \to M$ is a differentiable curve and $\alpha[\gamma]$ does not depend on the choice of representative. Both follow observing that

$$\begin{aligned}
\left.\frac{d}{dt}\right|_{t=0} (f \circ \gamma_1)(\alpha t) &= \sum_{j=1}^{k} \frac{\partial f}{\partial x^j} \left.\frac{d}{dt}\right|_{t=0} X^j \circ \gamma_1(\alpha t) \\
&= \sum_{j=1}^{k} \frac{\partial f}{\partial x^j} \frac{\alpha d}{d(\alpha t)} |_{t=0} X^j \circ \gamma_1(\alpha t) \\
&= \alpha \sum_{j=1}^{k} \frac{\partial f}{\partial x^j} \left.\frac{d}{dt}\right|_{t=0} X^j \circ \gamma_1(t) \\
&= \alpha \left.\frac{d}{dt}\right|_{t=0} f \circ \gamma_1(t) = \alpha \left.\frac{d}{dt}\right|_{t=0} f \circ \gamma_2(t) \\
&= \left.\frac{d}{dt}\right|_{t=0} f \circ \gamma_2(\alpha t)
\end{aligned}$$

for $\gamma_1, \gamma_2 \in [\gamma]$ and any differentiable $f : M \to \mathbb{R}$. In order to define addition, we first observe that a special representative can be found in the equivalence class of any curve $\gamma \in C_P$, which is given by

$$\gamma_s = \left[t \mapsto X^{-1}\left(t \left.\frac{d}{dt}\right|_{t=0} X \circ \gamma \right) \right],$$

which is defined in an open interval about the origin. Equivalence follows from

$$\begin{aligned}
\left.\frac{d}{dt}\right|_{t=0} f \circ \gamma &= \sum_{j=1}^{k} \frac{\partial f}{\partial x^j} \left.\frac{d}{dt}\right|_{t=0} X^j \circ \gamma \\
&= \left.\frac{d}{dt}\right|_{t=0} \left[f \circ X^{-1}\left(t \left.\frac{d}{dt}\right|_{t=0} X \circ \gamma \right) \right] \\
&= \left.\frac{d}{dt}\right|_{t=0} f \circ \gamma_s,
\end{aligned}$$

and independence on the choice of coordinates X is left as an exercise. We are now ready to define addition of equivalence classes of curves as

$$[\gamma_1] + [\gamma_2] = \left[X^{-1}\left(\cdot \left.\frac{d}{dt}\right|_{t=0} (X \circ \gamma_1 + X \circ \gamma_2) \right) \right].$$

This is made possible by the fact that $X \circ \gamma_j$, $j = 1, 2$, are curves in $\mathbb{R}^k$, so that the addition in the right-hand side makes sense. We only point out that the zero vector in $\mathcal{C}/\sim$ is given by the equivalence class of the curve $\gamma_P \equiv P$ and the opposite (additive inverse) of $[\gamma]$ is given by $[\gamma(-\cdot)]$. The verification that the definition is independent of the choice of representatives and of all vector space axioms are left as an exercise. Given a coordinate X about $P \in M$, it is an exercise to convince yourself that

$$[X^{-1}(\cdot\, e_j)], \quad j = 1, \dots, k,$$

is a basis of $T_P M$.

6.3.1 The differential

We conclude this chapter with a brief definition of a ubiquitous and fundamental object, that we all mindlessly use on a daily basis: the differential d. Let M^m and N^n be C^1-manifolds of the indicated dimensions $m, n \in \mathbb{N}$. We call a map $f : M \to N$ differentiable (or C^1, i. e., continuously differentiable) if

$$Y \circ f \circ X^{-1} : \mathbb{B}_{\mathbb{R}^m} \to \mathbb{B}_{\mathbb{R}^n} \text{ is differentiable (or } C^1\text{)}$$

for all charts X of the atlas of M and Y of the atlas of N, whenever X are coordinates for P and Y for $f(P)$ for any $P \in M$. The *differential* df of f is then given pointwise by

$$df_P : T_P M \to T_P N, [\gamma] \to [f \circ \gamma],$$

for $\gamma \in \mathcal{C}_P$, or, in coordinates,

$$d_P f(v) = \sum_{k=1}^{n} \sum_{j=1}^{m} v^j \frac{(\partial Y^k \circ f)}{\partial x^j} [Y^{-1}(\cdot\, e_k^n)] \in T_{f(P)} N,$$

for $v = \sum_{j=1}^{m} v^j [X^{-1}(\cdot\, e_j^m)] \in T_P(M)$. The superscripts in the basis vectors are used to indicate their different lengths. The standard notation used in books about topology or differential geometry are

$$f^k = Y^k \circ f, \quad \frac{\partial}{\partial x^j} = [X^{-1}(\cdot\, e_j^m)], \quad \text{and} \quad \frac{\partial}{\partial y^k} = [Y^{-1}(\cdot\, e_k^n)],$$

so that one can more compactly write

$$d_P f\left(\frac{\partial}{\partial x^j}\right) = \sum_{k=1}^{m} \frac{\partial f^k}{\partial x^j} \frac{\partial}{\partial y^k}, \quad j = 1, \dots, n.$$

Geometric objects (shapes), as we see them in nature or through our idealized imagination are simply regions of space and sets of points. In order to perform calculus with or on them, we need to use coordinates, much in the way the captain of a ship uses nautical charts. The geometric properties of shapes should, however, not depend on any specific coordinates used to derive them, since coordinates are a purely artificial device that is not intrinsic to the objects of study. For this reason, it is important to identify useful mathematical quantities and operations that are coordinate-free. The differential d is maybe the most fundamental such operator.

6.4 Concluding remarks

We can think of both submanifolds of $\mathbb{R}^d$ and abstract manifolds as sets M that are approximately linear in the vicinity of any of their points P. The linear structure is captured by the tangent spaces T_PM to the manifold at $P \in M$. Using local coordinates, we were able to understand smooth curves on M and smooth functions defined on M. More complicated maps, like for instance vector fields, i. e., maps $V : M \to TM$ with the property that

$$V(P) \in T_PM, \quad P \in M,$$

still pose a challenge. To see why, take a curve $\gamma : (-1, 1) \to M$ and study the behavior of V along γ, i. e., the map $V \circ \gamma$. Since the vectors

$$V(\gamma(t)) \in T_{\gamma(t)}M$$

are elements of different vector spaces, they cannot be added or subtracted, making it challenging to discuss their differentiability and their derivatives.[5] In the case of submanifolds in $\mathbb{R}^d$, all tangent spaces can be viewed as subspaces $T_PM \subset \mathbb{R}^d$ and the problem can be avoided. In a way, all tangent spaces are connected as subspaces of one and the same larger space $\mathbb{R}^d$. We can indeed consider the map $V \circ \gamma$ as a map from $(-1, 1)$ to $\mathbb{R}^d$. It can of course happen that $\frac{d}{dt}(V \circ \gamma)(t) \notin T_{\gamma(t)}M$. If we consider M to be our universe, however, we are only interested in the (rate of) change of direction (of the vector field) inside the tangent space. The latter can be obtained by projection onto the tangent space and leads to the definition

$$(\nabla_w V)(P) = \pi_{T_PM} \frac{d}{dt}(V \circ \gamma)(0),$$

where $\gamma(0) = P$, $w = \dot{\gamma}(0) \in T_PM$, and π_{T_PM} is the orthogonal projection from $\mathbb{R}^d$ to T_PM. Notice that, by choosing the orthogonal projection, we think of $\mathbb{R}^d$ as a Hilbert

5 Notice that TM can be given a natural manifold structure so that the smoothness of curves in TM can be understood in the way explained earlier in this chapter. The point we are making here, however, is another.

space with respect to its natural inner product, which naturally induces an inner product on each of the tangent spaces $T_P M$. Since the outcome only depends on $w \in T_P M$, we can replace w by a vector field W by using the interpretation

$$(\nabla_W V)(P) = (\nabla_{W(P)} V)(P)$$

Varying V and W while choosing γ so that $\dot{\gamma}(0) = W(P)$, it can be verified that $\nabla_W V$ is linear in W and V. It follows that, upon introduction of local coordinates about $P \in M$, it is enough to determine

$$\nabla_{\frac{\partial}{\partial x^j}} \frac{\partial}{\partial x^k} \in T_P M$$

for $i, j = 1, \dots, m$, where m is the dimension of the manifold, since the general case follows by linearity. Thus the operator ∇ is completely determined by the coefficients $\Gamma^i_{j,k} \in \mathbb{R}$ for which it holds

$$T_P M \ni \nabla_{\frac{\partial}{\partial x^j}} \frac{\partial}{\partial x^k} = \sum_{i=1}^{m} \Gamma^i_{j,k} \frac{\partial}{\partial x^i}.$$

The operator ∇ is called *connection* since it allows to connect the different tangent spaces as explained above. The connection depends on the choice of scalar product on $\mathbb{R}^d$ since it is defined using an orthogonal projection. When dealing with an abstract manifold M, we only have access to the tangent spaces $T_P M$ at points $P \in M$. It is therefore impossible to connect the tangent spaces in a "natural way" and one is free to choose any connection ∇. The choice of connection will determine the geometric properties of the manifold (which, without a connection, is merely a topological space, albeit with a differentiable structure). An important special case is that of Riemannian manifolds. These are manifolds, which have an inner product $\langle \cdot, \cdot \rangle_P$ defined for each tangent space $T_P M$ that depends smoothly on P. Such manifolds admit a "natural connection" ∇ (the so-called Levi-Civita connection) characterized by the validity of

$$d[\langle U, V \rangle](W) = \langle \nabla_W U, V \rangle + \langle U, \nabla_W V \rangle,$$

for any smooth vector fields U, V, W. When $M = \mathbb{R}^d$ with its natural scalar product, the corresponding Levi-Civita connection corresponds to D since

$$\begin{aligned} \partial_W (U \cdot V) &= d(U \cdot V)(W) = (DU)W \cdot V + U \cdot (DV)W \\ &= \partial_W U \cdot V + U \cdot \partial_W V, \end{aligned}$$

for smooth vector fields $U, V, W : \mathbb{R}^d \to \mathbb{R}^d$. Manifolds have many applications and are studied in topology and geometry.

7 Ordinary differential equations

Many a physical law can be written in terms of rates of change. The most basic and historically important example is that of Newton's law of motion

$$\frac{d}{dt}(mv) = ma = F,$$

which relates the motion of an object or the rate of change a (acceleration) of its speed v to the resultant F of all forces acting on it. If $x : [0, T) \to \mathbb{R}^d$ is the curve traced by the object up to time $T \leqslant \infty$, then this amounts to $m\ddot{x} = F$, where each dot indicates taking one derivative with respect to t. Considering that the force acting on the object may depend on its location, i. e., $F = F(x)$. We obtain an equation,

$$m\ddot{x} = F(x),$$

for its trajectory x through space. Think of planetary motion, where gravitational pull depends on relative position. This equation is an example of an *ordinary differential equation (ODE)*, i. e., an equation for an unknown function of one variable involving one or more of its derivatives. Another source of differential equations is biology. Consider the simplest population growth model where $x : [0, T) \to [0, \infty)$ is the size of a population as it changes over time. Assuming that the growth is proportional to the population size, by a factor $r > 0$, which is the growth rate, one arrives at the equation

$$\dot{x} = rx.$$

Borrowing from calculus knowledge, we may try and use the fundamental theorem of calculus to "undo" the derivative in a differential equation in order to find its solution. Taking the population growth example, we arrive at

$$x(t) = x_0 + \int_0^t rx(\tau)\, d\tau, \quad t \geqslant 0,$$

which makes it explicit that we need to know the initial size of the population in order to determine its subsequent evolution. This approach does, however, not produce a solution to the equation, but just another (integral) equation. In this formulation and fixing $T > 0$, we can think of the equation as a fixed-point equation for the unknown function $x : [0, T] \to \mathbb{R}$, which we may assume to be continuous. Indeed, if x is to be a solution of the ODE, it would even need to be differentiable. This means that we can think of x as an element of the vector space $C([0, T], \mathbb{R})$, which turns into a normed vector space if endowed with the supremum norm $\|\cdot\|_\infty$ given by

$$\|x\|_\infty = \sup_{t \in [0,T]} |x(t)|.$$

https://doi.org/10.1515/9783110780925-007

Defining

$$\Phi(x)(t) = x_0 + \int_0^t rx(\tau)\,d\tau, \quad t \in [0,T],$$

we obtain a map $\Phi : \mathrm{C}([0,T],\mathbb{R}) \to \mathrm{C}([0,T],\mathbb{R})$. Since $\Phi(x)(0) = x_0$, we can restrict this map to

$$X_0 = \mathrm{C}_{x_0}([0,T],\mathbb{R}) = \{x \in \mathrm{C}([0,T],\mathbb{R}) \mid x(0) = x_0\},$$

and obtain a self-map $\Phi : X_0 \to X_0$. Notice that

$$\begin{aligned}|\Phi(x)(t) - \Phi(y)(t)| &= \left| r\int_0^t (x(\tau) - y(\tau))\,d\tau \right| \leqslant r\int_0^t |x(\tau) - y(\tau)|\,d\tau \\ &\leqslant rT\|x - y\|_\infty, \quad t \in [0,T],\end{aligned}$$

which implies

$$\|\Phi(x) - \Phi(y)\|_\infty \leqslant rT\|x - y\|_\infty, \quad x, y \in X_0.$$

Considering that X_0 is a complete metric space (check this (Q1)) with respect to the metric induced by its norm, we can envision applying the Banach fixed-point theorem to obtain a solution. For that, we need that the map Φ be contractive. This can be achieved by choosing $T > 0$ so that $rT < 1$. In this way, we obtain a unique fixed point $x(\cdot, x_0) : [0, T_1] \to \mathbb{R}$ for $0 < T_1 < \frac{1}{r}$. As a fixed point of Φ, the function $x(\cdot, x_0)$ is differentiable by the fundamental theorem of calculus and satisfies

$$\dot{x}(t) = rx(t) \quad \text{for } t \in [0, T_1].$$

Then the continuity of $x(\cdot, x_0)$ implies that of $\dot{x}(\cdot, x_0)$, so that

$$x(\cdot, x_0) \in \mathrm{C}^1([0, T_1], \mathbb{R}).$$

At this point, we can consider the equation $\dot{x} = rx$ complemented with the new initial condition $x(0) = x_1 = x(T_1, x_0)$. The fixed-point argument did not rely on the specificity of the initial value x_0 and can be used to obtain a solution $x(\cdot, x_1)$ with the new initial datum x_1 on the interval $[0, T_1]$. Setting

$$x(t, x_0) = x(t - T_1, x_1), \quad t \in [T_1, 2T_1],$$

the solution $x(\cdot, x_0)$ can be extended to $[0, 2T_1]$, and, repeating this extension argument indefinitely, to $[0, \infty)$. We leave it to the reader to verify that this "piecewise" definition does not affect differentiability at the points $\{kT_1 \mid k \in \mathbb{N}\}$ of juncture. We shall see that this procedure can be used successfully also for more complicated (nonlinear) equations. For the simple equation $\dot{x} = rx$, the solution can of course be guessed. We, however, show another point of view that will inform our approach to general higher-dimensional linear ODEs. After rewriting the ODE as an integral equation, we can also keep replacing x by its integral representation to obtain

$$x(t) = x_0 + r\int_0^t x(\tau)\,d\tau = x_0 + x_0rt + r^2\int_0^t\int_0^\tau x(\sigma)\,d\sigma d\tau$$
$$= \left[1 + rt + \frac{1}{2}(rt)^2 + \cdots + \frac{1}{n!}(rt)^n\right]x_0 + r_n(t).$$

Noticing that $1 + rt + \frac{1}{2}(rt)^2 + \cdots + \frac{1}{n!}(rt)^n \nearrow e^{rt}$ as $n \to \infty$ and that

$$|r_n(t)| \leqslant \frac{(rT)^{n+1}}{(n+1)!}\|x\|_\infty, \quad t \in [0,T],$$

where the supremum norm is over the interval $[0,T]$, we first see that

$$\|x\|_\infty \leqslant e^{rT} + \alpha\|x\|_\infty$$

for $\alpha < 1$, if $n \in \mathbb{N}$ is chosen so that $\frac{(rT)^{n+1}}{(n+1)!} \leqslant \alpha < 1$, which is possible since $\frac{(rT)^{n+1}}{(n+1)!} \to 0$ as $n \to \infty$ for any fixed $T > 0$. It follows that $\|x\|_\infty \leqslant \frac{e^{rT}}{1-\alpha}$, so that x is bounded on $[0,T]$ and, therefore, that

$$r_n(t) \to 0 \text{ as } n \to \infty, \text{ uniformly in } t \in [0,T].$$

We conclude that $x(t) = e^{rt}x_0$ for $t \in [0,T]$ and for arbitrary $T > 0$, i. e., $x(t) = e^{rt}x_0$ for $t > 0$. By considering $y(t) = x(-t)$, we obtain similarly that $y(t) = e^{-rt}x_0$ for $t > 0$, so that $x(t) = e^{rt}x_0$ is a solution also for $t < 0$.

7.1 Linear systems of ODEs

An important class of ODEs is that of linear equations. We now consider trajectories/curves x in $\mathbb{R}^n$, i. e.,

$$x : [0,T] \to \mathbb{R}^n,$$

or, more in general, at no additional effort, curves $x : [0,T] \to E$ in a complete normed vector space E (which we called Banach space), possibly infinite-dimensional. Starting with $\mathbb{R}^n$, a linear system of ODEs is simply an equation such as

$$\dot{x} = Ax, \quad x(0) = x_0,$$

for an unknown curve $x : [0, T] \to \mathbb{R}^n$ and a matrix $A \in \mathbb{R}^{n \times n}$. More in general one would consider an ODE like

$$\dot{x} = F(x), \quad x(0) = x_0,$$

where $F : \mathbb{R}^n \to \mathbb{R}^n$ is a nonlinear function. Thus one speaks of linear equation when F is linear. As to the importance of linear equations, we observe that a solution to the nonlinear equation, if it exists, will stay close to x_0 for some possibly small interval of time. If F is differentiable, then

$$F(x) \simeq F(x_0) + Df(x_0)(x - x_0)$$

and $y = x - x_0$ solves $\dot{y} = Ay + F(x_0)$, $y(0) = 0$. This is essentially a linear equation, modulo the additional constant term, and is the reason why we shall consider the slightly more general inhomogeneous equation

$$\dot{x} = Ax + f(t), \quad x(0) = x_0,$$

for $f : [0, T] \to \mathbb{R}^n$. As an example of a linear homogeneous ($f \equiv 0$) system of ODEs consider two biological species x_1 and x_2 living in the same environment and evolving according to the simple rule that they each grow at a rate r_1 and r_2, respectively, and that one preys on the other so that

$$\begin{cases} \dot{x}_1 = r_1 x_1 + \alpha x_2, & x_1(0) = x_{1,0}, \\ \dot{x}_2 = r_2 x_2 - \beta x_1, & x_2(0) = x_{2,0}, \end{cases}$$

for parameters $\alpha, \beta > 0$ modeling the enhanced growth experienced by the predator population, which depends on the availability of prey, and the reduced growth of the prey population, influenced by the size of the predator population.

Integrating the linear homogeneous equation yields

$$x(t) = x_0 + \int_0^t Ax(\tau)\, d\tau,$$

and, iterating the procedure, leads to

$$x(t) = x_0 + tAx_0 + \frac{t^2 A^2}{2} x_0 + \cdots + \frac{t^n A^n}{n!} x_0 + r_n(t),$$
$$r_n(t) = A^{n+1} \int_0^t \int_0^{\tau_1} \cdots \int_0^{\tau_{n-1}} x(\sigma)\, d\sigma d\tau_{n-1} \ldots d\tau_1$$

When working in a Banach space $(E, \|\cdot\|_E)$, A can be taken to be a bounded[1] linear operator, i. e., a linear map $A : E \to E$ satisfying

$$\|A\| = \sup_{\|x\|_E=1} \|Ax\|_E = \sup_{x\neq 0} \frac{\|Ax\|_E}{\|x\|_E} < \infty.$$

It is an exercise to prove that $\|A^2\| \leqslant \|A\|^2$, and consequently, that one has $\|A^n\| \leqslant \|A\|^n$ for any $n \geqslant 3$ also, for a bounded operator $A \in \mathcal{L}(E,E)$, where

$$\mathcal{L}(E,E) = \{A : E \to E \mid A \text{ linear and } \|A\| < \infty\}.$$

With this in hand, we can show that $s_n = \sum_{k=0}^{n} \frac{(tA)^k}{k!}$ converges as $n \to \infty$. Indeed, it holds that

$$\|s_n - s_m\| = \left\| \sum_{k=m+1}^{n} \frac{(tA)^k}{k!} \right\| \leqslant \sum_{k=m+1}^{n} \frac{t^k \|A\|^k}{k!} \longrightarrow 0 \quad \text{as } m, n \to \infty,$$

since $(\sum_{k=0}^{n} \frac{t^k\|A\|^k}{k!})_{n\geqslant 0}$ converges to $e^{t\|A\|}$, and thus is a Cauchy sequence. This is true for any $t > 0$ and the convergence is uniform for $t \in [0, T]$. Thus, if the space $(\mathcal{L}(E,E), \|\cdot\|)$ is complete, a limit of the Cauchy sequence $(s_n)_{n\in\mathbb{N}}$ must exist. We use the notation

$$e^{tA} = \sum_{k=0}^{\infty} \frac{t^k A^k}{k!},$$

and observe (without proof) that

$$\frac{d}{dt} e^{tA} = \frac{d}{dt} \sum_{k=0}^{\infty} \frac{t^k A^k}{k!} = \sum_{k=0}^{\infty} \frac{d}{dt} \frac{t^k A^k}{k!} = A \sum_{k=1}^{\infty} \frac{t^{k-1} A^{k-1}}{(k-1)!} = Ae^{tA}, \quad t > 0.$$

This exponential function for linear bounded operators shares some properties with the real (or complex) exponential function. There are, however, differences. Show, for instance, that $e^{A+B} \neq e^A e^B$, in general, even for $A, B \in \mathbb{R}^{2\times 2}$. Can you identify the origin of the failure of this identity?

We leave it as an exercise to check that $r_n(t) \to 0$ as $n \to \infty$ for any $t > 0$ and, uniformly in $t \in [0, T]$ for any $T > 0$.

Fact. *The vector space $\mathcal{L}(E,E)$ is complete with respect to the norm $\|\cdot\|$.*

---

1 Boundedness in this context is equivalent to continuity of A. The reader may try to prove this equivalence when E is finite-dimensional, where all linear operators can be shown to be continuous/bounded.

Let $(L_n)_{n\in\mathbb{N}}$ be a Cauchy sequence of bounded linear operators, i. e., satisfying

$$\forall \varepsilon > 0\ \exists N \in \mathbb{N} \quad \text{with } \|L_n - L_m\| \leqslant \varepsilon \text{ for } m, n \geqslant N.$$

Then, given $x \in E$, we can find $N \in \mathbb{N}$ such that $\|L_n - L_m\| \leqslant \frac{\varepsilon}{\max\{\|x\|_E, 1\}}$, and thus

$$\|L_n x - L_m x\|_E \leqslant \|L_n - L_m\| \|x\|_E \leqslant \frac{\varepsilon}{\max\{\|x\|_E, 1\}} \|x\|_E \leqslant \varepsilon,$$

for $m, n \geqslant N$. This shows that, for each $x \in E$, $(Lx_n)_{n\in\mathbb{N}}$ is a Cauchy sequence in E, which has a limit $L_\infty(x)$, since E is assumed complete. This defines a map $L_\infty : E \to E$, which is linear since

$$\begin{aligned} L_\infty(\lambda x + y) &= \lim_{n\to\infty} L_n(\lambda x + y) = \lim_{n\to\infty} (\lambda L_n x + L_n y) \\ &= \lambda \lim_{n\to\infty} L_n x + \lim_{n\to\infty} L_n y = \lambda L_\infty(x) + L_\infty(y), \end{aligned}$$

for $x, y \in E$, $\lambda \in \mathbb{R}$. Next, observe that the triangle inequality gives (Q2)

$$\big|\|L_n\| - \|L_m\|\big| \leqslant \|L_n - L_m\|, \quad m, n \in \mathbb{N},$$

which shows that $(\|L_n\|_n)_{n\in\mathbb{N}}$ is a Cauchy sequence in $\mathbb{R}$ and, therefore, has a limit $M \geqslant 0$. It follows that

$$\|L_n x\|_E \leqslant \|L_n\| \|x\|_E \leqslant (M+1)\|x\|_E,$$

for $n \geqslant N$ and some $N \in \mathbb{N}$. Letting n tend to ∞ in $L_n x$ yields

$$\|L_\infty x\| \leqslant (M+1)\|x\|_E,$$

which amounts to the boundedness of L_∞. √

We have effectively shown that the linear equation

$$\dot{x} = Ax, \quad x(0) = x_0 \in E,$$

possesses the solution $x(t) = e^{tA}x_0$, $t \in \mathbb{R}$, for any $x_0 \in E$ and any linear bounded operator A. How (Q3) would you show that this solution is unique? How do you obtain this solution for $t < 0$?

In order to avoid introducing the integral of Banach space valued continuous (or more general) functions $f : [0, T] \to E$, which is possible and goes by the name of *Bochner integral*, we restrict our attention, for the remainder of this section, to the finite-dimensional case, which further reduces to $\mathbb{R}^n$ upon introduction of coordinates. Given $T > 0$ and $f \in C([0, T], \mathbb{R}^n)$, the inhomogeneous linear equation

$$\dot{x} = Ax + f(t), \quad x(0) = x_0,$$

can be solved with the use of $e^{\cdot A}$ only, in a way very much similar to that you have likely learned in the one-dimensional case in a first course about ODEs. The basic idea is that of using an integrating factor, in this case the exponential of the matrix. Setting $y = e^{-tA}x$, we see that

$$\dot{y} = \frac{d}{dt}(e^{-tA}x) = -Ae^{-tA}x + e^{-tA}\dot{x} = e^{-tA}f(t), \quad t \in [0, T],$$

using the "product rule" and that $Ae^{-tA} = e^{-tA}A$ for $t \in \mathbb{R}$, facts the validity of which you are encouraged to verify on your own. Integrating (component by component) then yields

$$e^{-tA}x(t) = y(t) = x_0 + \int_0^t e^{-\tau A}f(\tau)\,d\tau, \quad t \in [0, T],$$

and, upon verification that $(e^{tA})^{-1} = e^{-tA}$ for $t \in \mathbb{R}$ (left to the reader), the formula

$$x(t) = e^{tA}x_0 + \int_0^t e^{(t-\tau)A}f(\tau)\,d\tau, \quad t \in [0, t].$$

This formula is known as the *variation of constants formula*.[2]

While refraining at this stage from going into a detailed discussion, we observe that, while e^{tA} is well-defined, we do not have much of an idea of how it actually looks like for any concrete example. The defining series is not much help in an actual computation of the exponential. We merely notice that, if it is possible to find a vector $v \neq 0$ such that $Av = \lambda v$ for some $\lambda \in \mathbb{C}$ (in this case λ is called an *eigenvalue* and v an *eigenvector* of A), then

$$A^k v = \lambda^k v, \quad k \in \mathbb{N}$$

and, therefore,

$$e^{tA}v = \sum_{k=0}^{\infty} \frac{(tA)^k}{k!}v = \sum_{k=0}^{\infty} \frac{t^k}{k!}A^k v = \sum_{k=0}^{\infty} \frac{(t\lambda)^k}{k!}v = e^{\lambda t}v,$$

for $t \in \mathbb{R}$. If we had a whole basis $v_1, \dots, v_n$ for $\mathbb{R}^n$ comprised of eigenvectors (to eigenvalues $\lambda_1, \dots, \lambda_n$), we could write

$$x_0 = \sum_{j=1}^{n} x_0^j v_j,$$

2 In the context of partial differential equations, its generalization is also known as *Duhamel's principle*.

for coefficients x_0^j, $j = 1, \dots, n$, in this basis. Then

$$e^{tA}x_0 = \sum_{j=1}^{n} x_0^j e^{\lambda_j t} v_j,$$

greatly simplifying the computation of the exponential. While it is not possible to find a basis consisting solely of eigenvectors for general matrices, we shall show in Chapter 8 that this is possible for symmetric matrices, i. e., matrices A with $A^\top = A$.

7.2 Existence and uniqueness for nonlinear ODEs

Most differential equations are nonlinear and do not admit explicit representations for their solutions. It is therefore important to be able to determine existence, uniqueness, and long time behavior based on the analysis of the equations alone. These questions are of interest even if it is possible to use numerical methods to compute solutions. Numerical procedures can in fact fail to deliver an approximation to the correct solution or to converge at all. Often numerical issues can be addressed if their origin is understood, thus providing additional motivation for the analysis of ODEs. Their analysis plays an important role in differential geometry as well, where basic objects such as geodesics (curves of minimal distance between points) appear as solutions of nonlinear systems of ODEs. Here, we only focus on the basic existence and uniqueness result for nonlinear ODEs, which we view as an application of the Banach fixed-point theorem of Section 5.5.1.

We first consider simple examples to illustrate that obstacles and limitations are encountered when studying the existence and uniqueness for ODEs. The nonlinear scalar equation

$$\dot{x} = x^2, \quad x(0) = x_0 > 0,$$

has the unique (will follow from the upcoming discussion) solution given by $x(t) = \frac{x_0}{1-x_0 t}$, which is seen to develop a singularity at $t = \frac{1}{x_0}$. This shows that solutions of nonlinear ODEs may not exist for all times $t \in \mathbb{R}$, while they actually do in the linear case as shown in the previous section. Both functions $x \equiv 0$ and $x(t) = t^3$, $t \in \mathbb{R}$ are solutions of

$$\dot{x} = 3x^{\frac{2}{3}}, \quad x(0) = 0,$$

showing that the equation admits multiple solutions. The condition that allows the construction of a unique solution turns out to be Lipschitz continuity. For a function $F : E \to E$ on a Banach space E with norm $|\cdot|_E$, *Lipschitz continuity* amounts to the existence of a constant $L \geqslant 0$ such that

$$|F(x) - F(y)|_E \leqslant L|x - y|_E, \quad x, y \in E.$$

We speak in this case of *global* Lipschitz continuity in that the constant L works for all $x, y \in E$. More often, it is only possible and enough for our purposes that Lipschitz continuity be available only *locally*. A function $F : E \to E$ is called *locally Lipschitz continuous* iff

$$\forall x \in E\, \exists r = r(x) > 0,\ \exists L = L(x) \geqslant 0 \text{ with } |F(y) - F(z)|_E \leqslant L|y - z|_E, \quad y, z \in \mathbb{B}_E(x, r).$$

This simply means that Lipschitz continuity holds in a neighborhood of any and each $x \in E$. Notice that $F(x) = x^2$, $x \in \mathbb{R}$ is an example of a locally, but not globally, Lipschitz function, whereas $F(x) = 3x^{\frac{2}{3}}$, $x \in \mathbb{R}$ is not Lipschitz in any neighborhood of $x = 0$.

Fact. *Assume that $F : \mathbb{R}^n \to \mathbb{R}^n$ be locally Lipschitz. Then, given $x_0 \in \mathbb{R}^n$, there is a unique solution $x(\cdot, x_0)$ of the initial value problem*

$$\dot{x} = F(x), \quad x(0) = x_0,$$

on the interval $[0, T]$ for some $T > 0$. This solution can be extended to a maximal interval of existence $[0, T(x_0))$ with $T(x_0) > 0$. If $T(x_0) < \infty$, then

$$\lim_{t \nearrow T(x_0)} |x(t, x_0)|_2 = \infty.$$

We shall use the Banach fixed-point theorem as we did for the linear equation. For fixed $T > 0$ (to be chosen later), define

$$X = X(T, r, x_0) = \{x \in \mathrm{C}([0, T], \mathbb{R}^n) \mid x(0) = x_0,\ |x(t) - x_0|_2 \leqslant r,\ t \in [0, T]\},$$

which is a complete subset of $\mathrm{C}([0, T], \overline{\mathbb{B}}(x_0, r))$ with respect to the supremum norm $\|\cdot\|_\infty$. Define

$$\Phi(x)(t) = x_0 + \int_0^t F(x(\tau))\, d\tau, \quad t \in [0, T],\ x \in X,$$

and observe that

$$|\Phi(x)(t) - x_0|_2 \leqslant \int_0^t |F(x(\tau))|_2\, d\tau, \quad t \in [0, T],\ x \in X.$$

Since continuous maps assume their maximum and minimum on compact sets, we have that

$$|F(z)|_2 \leqslant M, \quad z \in \overline{\mathbb{B}}(x_0, r),$$

for some $M \geqslant 0$, and consequently, that

$$|\Phi(x)(t) - x_0|_2 \leqslant MT, \quad t \in [0,T],\ x \in X.$$

By choosing T sufficiently small, we can achieve that $\Phi(x)(t) \in \overline{\mathbb{B}}(x_0, r)$ for $t \in [0,T]$, which shows that Φ maps X into itself. Notice that this is possible for any $r > 0$. Local Lipschitz continuity yields $r_0 > 0$ and $L \geqslant 0$ such that

$$|F(y) - F(z)|_2 \leqslant L\,|y - z|_2, \quad y, z \in \overline{\mathbb{B}}(x_0, r_0).$$

Then it holds that

$$\begin{aligned} |\Phi(x)(t) - \Phi(\tilde{x})(t)|_2 &\leqslant \int_0^T |F(x(\tau)) - F(\tilde{x}(\tau))|_2\, d\tau \\ &\leqslant LT\|x - \tilde{x}\|_\infty. \end{aligned}$$

This shows that

$$\|\Phi(x) - \Phi(\tilde{x})\|_\infty \leqslant \frac{1}{2}\|x - \tilde{x}\|_\infty, \quad x, \tilde{x} \in X,$$

provided $T \leqslant T_0$ and T_0 is chosen so that $LT_0 \leqslant \frac{1}{2}$. The Banach fixed-point theorem then yields a unique solution $x(\cdot, x_0) \in X = X(T_0, r_0, x_0)$. The reader can verify that such a fixed point is a solution of the initial value problem and that any solution of the initial value problem must be a fixed point of Φ. The length of the existence interval $[0, T_0]$ depends on the local properties of F near the point x_0. Considering the equation

$$\dot{x} = F(x), \quad x(0) = x_1 = x(T_0, x_0),$$

the argument can be repeated to obtain a solution $x(\cdot, x_1)$ on some interval $[0, T_1]$ for $T_1 > 0$. Glueing the solutions together, i. e., extending $x(\cdot, x_0)$ by setting

$$x(t, x_0) = x(t - T_0, x_1), \quad t \in [T_0, T_0 + T_1],$$

we obtain a (still unique) solution on $[0, T_0 + T_1]$, and, repeating the argument, on an interval $[0, T(x_0))$ of length $T(x_0) = \sum_{k \geqslant 0} T_k$, where $T_k > 0$ for $k \geqslant 0$. Clearly, we either have $T(x_0) = \infty$ or $T(x_0) < \infty$. In the latter case, if $\limsup_{t \nearrow T(x_0)} |x(t, x_0)|_2 < \infty$,[3] there exists a sequence $(t_k)_{k \in \mathbb{N}}$ with $t_k \nearrow T(x_0)$ as $k \to \infty$ such that $x(t_k, x_0) \to x_\infty \in \mathbb{R}^n$, in which case the solution could be extended once more using x_∞ as a new initial datum.

3 The *limes superior* $\limsup_{k \to \infty} x_k$ *of a sequence* $(x_k)_{k \in \mathbb{N}}$ of reals is defined as $\lim_{n \to \infty} \bar{x}_n$, where $(\bar{x}_n)_{n \in \mathbb{N}}$ is the nonincreasing sequence given by $\bar{x}_n = \sup_{k \geqslant n} x_k$, $n \in \mathbb{N}$. The limit can be infinite. Derive the definition for a function (as opposed to for a sequence), which is needed here.

This argument also shows that the maximal interval of existence cannot include the right-end point. √

If F is globally Lipschitz, as is the case for F linear (affine), the above argument shows that we can choose $T_k = T_0$ for $k \geqslant 1$ and obtain a global solution, i. e. one for which $T(x_0) = \infty$ for any x_0. The examples in the beginning of the section show that global existence does not hold in general if the global Lipschitz property is dropped and that uniqueness is lost, in general, if no Lipschitz continuity is required.

7.3 Gradient flows

A very important special class of ODEs is obtained when the driving vector field $F : \mathbb{R}^n \to \mathbb{R}^n$ ($n \in \mathbb{N}$) has a special form. This happens when there is a function $u : \mathbb{R}^n \to \mathbb{R}$ with

$$-\nabla u(x) = F(x), \quad x \in \mathbb{R}^n,$$

where $\nabla u(x) = Du(x)^\top \in \mathbb{R}^{n\times 1} = \mathbb{R}^n$. The corresponding ODE system reads

$$\dot{x} = -\nabla u(x), \quad x(0) = x_0,$$

and, geometrically, it pushes a solution going through a point $z \in \mathbb{R}^n$ in the direction of steepest descent of the values of u about z. This can be seen by using the Cauchy–Schwarz inequality to obtain that

$$\partial_v u(z) = Du(z)v = \nabla u(z)^\top v \geqslant -|\nabla u(z)|_2 |v|_2 = -|\nabla u(z)|_2,$$

for $v \in \mathbb{S}^{n-1}$ and observing that the strongest decrease happens in the direction of the vector $v = -\frac{\nabla u(z)}{|\nabla u(z)|_2}$. If the gradient vanishes at the point z, then $x \equiv z$ is a solution of the initial value problem with initial datum z. We shall assume that $u \in C^1(\mathbb{R}^n, \mathbb{R})$ and that ∇u is Lipschitz (locally) in order to ensure unique existence of solutions to the initial value problem. Now, if $x(\cdot)$ is a solution of the initial value problem, it satisfies

$$\begin{aligned}\frac{d}{dt}(u \circ x)(t) &= Du(x(t))\dot{x}(t) = -\nabla u(x(t))^\top \nabla u(x(t))\\ &= -|\nabla u(x(t))|_2^2 \leqslant 0\end{aligned}$$

for as long as it exists. This shows that, if $u : \mathbb{R}^n \to \mathbb{R}$ is coercive in the sense that

$$\lim_{|z|_2 \to \infty} u(z) = \infty,$$

then any solution exists globally. If it did not, it would hold that

$$\lim_{t \nearrow T(x_0)} |x(t, x_0)|_2 = \infty,$$

for some $x_0 \in \mathbb{R}^n$ and $T(x_0) < \infty$. This is, however, excluded by the validity of

$$u(x(t, x_0)) = u(x_0) + \int_0^t \frac{d}{dt}(u(x(\tau, x_0)))(\tau)\, d\tau \leqslant u(x_0),$$

and the fact that u is bounded from below, i. e., $u \geqslant M$ for some $M \in \mathbb{R}$, as follows from its coercivity (why? (Q4)). This also shows that solutions are bounded. Since closed and bounded subsets of the real line are (sequentially) compact, we can find $(t_k)_{k \in \mathbb{N}}$ with $t_k \to \infty$ for $k \to \infty$ such that

$$x(t_k, x_0) \to x_\infty \quad \text{as } k \to \infty,$$

for some $x_\infty \in \mathbb{R}^n$. If we assume that ∇u is bounded on bounded sets, i. e., if it holds that

$$\forall R > 0\; \exists M = M(R) \geqslant 0 \quad \text{with } |\nabla u(z)|_2 \leqslant M \text{ for } |z|_2 \leqslant R,$$

then we can show that it must necessarily hold that $\nabla u(x_\infty) = 0$. This means that $x(\cdot, x_\infty) \equiv x_\infty$ is a steady state (equilibrium) for the initial value problem. We argue by contradiction and assume that $\nabla u(x_\infty) \neq 0$. Boundedness of $x(\cdot, x_0)$ yields $R > 0$ such that $|x(\cdot, x_0)|_2 \leqslant R$, and thus that

$$|\dot{x}(t)|_2 = |\nabla u(x(t, x_0))|_2 \leqslant M,$$

for some $M \geqslant 0$ and $t \geqslant 0$. Then we have that

$$|x(t, x_0) - x(s, x_0)|_2 = \left| \int_s^t \dot{x}(\tau, x_0)\, d\tau \right|_2 \leqslant M|t - s|, \quad t, s \geqslant 0.$$

Continuity of ∇u ensures the existence of $r_\infty > 0$ such that

$$|\nabla u(z)|_2 \geqslant \frac{|\nabla u(x_\infty)|_2}{2} > 0 \quad \text{for } z \text{ with } |z - x_\infty| \leqslant r_\infty.$$

Then

$$|x(\tau, x_0) - x(t_k, x_0)|_2 \leqslant r_\infty/2 \quad \text{for } \tau \text{ with } |t_k - \tau| \leqslant \delta,$$

where $\delta = r_\infty/2M > 0$. There is also $K \in \mathbb{N}$ with $|x(t_k, x_0) - x_\infty|_2 \leqslant r_\infty/2$ for $k \geqslant K$. It then follows that

$$
\begin{aligned}
u(x(t,x_0)) &\leqslant u(x_0) - \int_0^t |\nabla u(x(\tau,x_0))|_2^2 \, d\tau \\
&\leqslant u(x_0) - \sum_{k\geqslant K,\, t_k\leqslant t} \int_{t_k-\delta}^{t_k} |\nabla u(x(\tau,x_0))|_2^2 \, d\tau \\
&\leqslant u(x_\infty) - \frac{|\nabla u(x_\infty)|_2^2}{4} \sum_{k\geqslant K,\, t_k\leqslant t} \delta,
\end{aligned}
$$

which yields a contradiction as $t \to \infty$ since $\sum_{k\geqslant K,\, t_k\leqslant t} \delta \to \infty$.

This shows that the limit points of solutions $x(\cdot, x_0)$ of a gradient system as $t \to \infty$, with the above assumptions, are critical points of the potential u, i. e., points where ∇u vanishes. We will see an application of this in the next chapter.

7.4 Concluding remarks

The study of the behavior of solutions to ordinary differential equations (and more general equations) goes by the name of dynamical systems. The latter studies the totality of all solutions of an equation and aims at establishing its structural properties. Of particular interest are often the existence of special solutions (like equilibria and periodic solutions) and their stability, as well as the existence and properties of special sets (invariant manifolds, attractors, etc.). Ordinary differential equations can also be considered on a manifold M. They take the form

$$
\begin{cases}
\dot{X} = F(X), \\
X(0) = P,
\end{cases}
$$

for a smooth vector field $F : M \to TM, P \mapsto F(P) \in T_P M$. A solution on $[0, T)$ is a curve $\gamma : [0, T) \to M$, which satisfies $\gamma(0) = P$ and $\dot{\gamma}(t) = F(\gamma(t))$ for $t \in [0, T)$.[4] Notice that, upon introduction of local coordinates on the manifold, the equation reduces (Q5) to an ordinary differential equation in $\mathbb{R}^m$, where m is the dimension of the manifold, at least in small intervals of time. Indeed the solution may leave any chosen initial coordinate patch containing the initial point P, so that different local coordinates need to be used in different time segments. The existence result we described earlier in the chapter applies to the manifold setting since it is local ("stepwise") in nature.

---

4 We use an intuitive notation here instead of the more correct form $d_t\gamma(\frac{\partial}{\partial t}) = F(\gamma(t))$ introduced in Section 6.3.

8 Optimization

Optimization problems are ubiquitous in all human endeavors: mathematics, science, engineering, computer science, business, health care, etc. We like to measure things and like to see “good numbers”. This means that we often introduce a metric to measure benefit or harm, which depends on variables we have control over and look for those values of the variables that maximize the benefit or minimize the harm. In geometry, we may want to identify which shape contains the most volume given its surface. In physics, one is often interested in configurations that minimize some kind of energy. In business, profit is typically the goal of optimization. For Netflix, it may be the likelihood of you liking a certain type of content, so they can offer the best suggestions in order to keep you engaged. Mathematically, this boils down to having an objective function $f : X \to \mathbb{R}$ and looking for its extremal points. If X is only a set with no additional structure, the search can only occur through a scan of all individual arguments $x \in X$ in order to identify which ones lead to the extreme values of f, if they exist. If the set X has additional structure, it is possible to devise some more sophisticated strategies. Here, we focus mainly on the continuous case (as opposed to discrete) and assume that X is either a linear space or a smooth manifold. The dimension of X could be infinite as we shall see in the final remarks of the next chapter. We start with an objective function f with $f \in C^1(\mathbb{R}^d, \mathbb{R})$ for some $d \in \mathbb{N}$, which we would like to minimize in the sense that we look for the argument(s) such that f has the least possible value

$$\operatorname{argmin}_{x \in X} f(x) = \operatorname{argmin}_X f = \{y \in X \mid f(y) \leq f(x)\ \forall\, x \in X\}$$

With no information in hand, we begin with some random initial guess $x_0 \in \mathbb{R}^d$ and test our way around for a direction $v \in \mathbb{R}^d$ along which f decreases. We know that the direction of steepest descent at x_0 is given by

$$v = -\frac{\nabla f(x_0)}{|\nabla f(x_0)|_2},$$

something we learned toward the end of the previous chapter. This naturally leads us to consider the initial value problem

$$\dot{x} = -\alpha \nabla f(x), \quad x(0) = x_0,$$

which may generate a solution that flows toward a minimizer. The intensity $\alpha > 0$ with we which we take off in direction of the negative gradient is a parameter that may need to be tuned in concrete applications (and even varied in time). In practical implementations, continuous time is not viable and a natural replacement is time stepping. It can be obtained by taking discrete time steps of size h and approximating the time

https://doi.org/10.1515/9783110780925-008

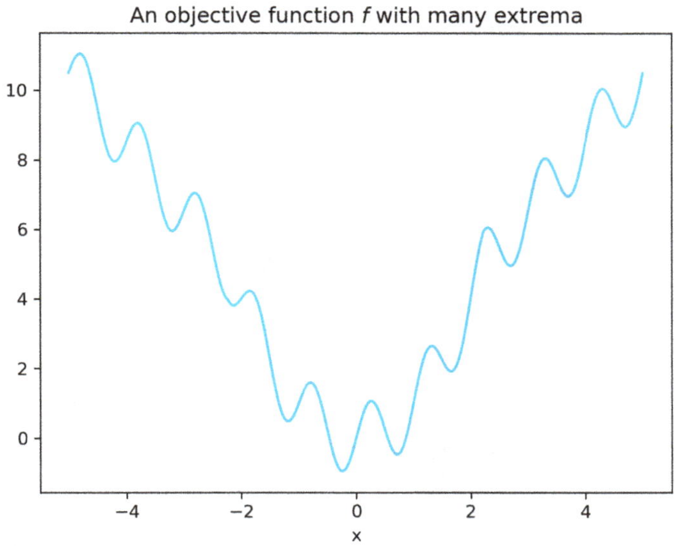

Figure 8.1: An objective function with multiple wells.

derivative by a difference quotient

$$\frac{x(t_n + h) - x(t_n)}{h} = -\alpha \nabla f(x(t_n)), \quad n \geqslant 0,\ x(t_0) = x_0.$$

Taking $t_n = t_0 + nh$ for $n \geqslant 0$ and denoting $x(t_n)$ by x_n, this amounts to the recursion relation

$$x_{n+1} = x_n - \alpha h \nabla f(x_n), \quad n \geqslant 0,\ x_0 \text{ given.}$$

In other words, we take steps in the direction of steepest decrease from where we happen to be and keep repeating this procedure indefinitely. This iteration is by no means guaranteed to succeed: the function f maybe unbounded from below as for $f(x) = -x^2$, in which case we would wander off to infinity (check). The function f may have multiple extremal points like the one the graph of which is depicted in Figure 8.1 and our walk may lead us to a local but not global minimum, depending on where we start our search. In spite of this, the basic idea of gradient descent is a very commonly used tool in optimization, albeit often in some modified form. A particularly well-suited class that allows for straightforward minimization is that of convex functions, of which quadratic functions are the model example. We will get back to convexity after considering one simple in this section and two more involved examples in the next two sections. The simple one is the objective function given by

$$f(x) = \frac{1}{2}|x|_2^2, \quad x \in \mathbb{R}^d,$$

the gradient flow of which is simply given by $\dot{x} = -\alpha x$. Starting with the initial guess x_0, the solution of the ODE system is given by $x(t) = e^{-\alpha t}x_0$. It converges to the unique point of minimum $x_\infty = 0$ at a speed that is determined by the parameter $\alpha > 0$. Taking the discrete approach, we would consider the recursion

$$x_{n+1} = x_n - \alpha\nabla f(x_n) = x_n - \alpha x_n, \quad x_0 \text{ given.}$$

The parameter $\alpha > 0$ (in which we absorbed the time step h of above) could be chosen by looking for the minimum of f along the line $l = [\alpha \to x_n - \alpha x_n]$, i. e.,

$$\text{argmin}_{\alpha>0} f \circ l = \text{argmin}_{\alpha>0} \frac{1}{2}|(1-\alpha)x_0|_2^2 = \frac{1}{2}|x_0|_2^2 \,\text{argmin}_{\alpha>0}(1-\alpha)^2$$

in the first step $n = 0$, which yields $\alpha = 1$ and the actual global minimum is reached in one single step since $x_1 = 0$, in this case. This is of course due to the special radial nature of the function f.

8.1 Singular value decomposition

We consider now matrices $A \in \mathbb{R}^{n\times m}$ and try to understand their structure. Using simple matrices as building blocks, we try to recover any matrix as a linear combination of these. By a simple matrix in $\mathbb{R}^{n\times m}$, we mean one that maps the whole domain $\mathbb{R}^m$ to a one-dimensional subspace of $\mathbb{R}^n$. The zero matrix would be even simpler but not very useful. A matrix A with one-dimensional range is what we called a *rank one* matrix and can be written as

$$A = \sigma v u^\top,$$

for unit length vectors $0 \neq u \in \mathbb{R}^m$ and $0 \neq v \in \mathbb{R}^n$ and a scalar $\sigma \neq 0$. In order to see this, observe that the columns $A_i^\bullet \in \mathbb{R}^n$, $i = 1, \dots, m$, must all be multiples $A_i^\bullet = \tilde{u}_i\tilde{v}$ of a single vector $0 \neq \tilde{v} \in \mathbb{R}^n$ with $\mathbb{R} \ni \tilde{u}_i \neq 0$ for at least one index since $A \neq 0$. It follows that

$$A = \begin{bmatrix} \tilde{u}_1\tilde{v} & \tilde{u}_2\tilde{v} & \cdots & \tilde{u}_n\tilde{v} \end{bmatrix} = \tilde{v}\tilde{u}^\top.$$

Since the vectors $\tilde{u}, \tilde{v}$ are nonzero, we see that

$$\tilde{v}\tilde{u}^\top = |\tilde{v}|_2|\tilde{u}|_2 \frac{\tilde{v}}{|\tilde{v}|_2}\frac{\tilde{u}^\top}{|\tilde{u}|_2} = \sigma v u^\top,$$

where u, v have unit length. Think of this matrix as taking any vector $x \in \mathbb{R}^m$ and computing its component $u^\top x$ in the direction of u (u can be thought of as the first vector in an orthonormal basis of $\mathbb{R}^m$) and producing a multiple of the vector $v \in \mathbb{R}^n$ with factor $\sigma u^\top x$. The computation of the component of x in direction of u can be interpreted as a projection and σ as a stretching factor. If a matrix $A \in \mathbb{R}^{n\times m}$ has rank $k \leqslant \min\{m, n\}$, then it is conceivable that it be representable as

$$A = \sum_{j=1}^{k} \sigma_j v_j u_j^\top,$$

for so-called *singular values* $\sigma_j \in (0, \infty)$ and linearly independent unit vectors $u_j \in \mathbb{R}^m$ and $v_j \in \mathbb{R}^n$. We shall see that this is possible and that the vectors can be chosen so that $u_i^\top u_j = \delta_{ij}$ that $v_i^\top v_j = \delta_{ij}$ for $i, j = 1, \ldots, k$, respectively.

The space $\mathbb{R}^{n\times m}$ of matrices is a vector space with respect to matrix addition and scalar multiplication. Of the several norms that can be used on it, we now select the so-called *Frobenius norm* $\|\cdot\|_2$, which is defined through

$$\|A\|_2^2 = \sum_{j=1}^{n}\sum_{i=1}^{m} |A_i^j|^2 = \operatorname{Tr}(A^\top A) = \operatorname{Tr}(AA^\top), \quad A \in \mathbb{R}^{n\times m},$$

where the trace of a square matrix $M \in \mathbb{R}^{d\times d}$, $d \in \mathbb{N}$, is given by

$$\operatorname{Tr}(M) = \sum_{k=1}^{d} M_k^k.$$

In search of singular values, we look for the "best" rank one approximation $\sigma_1 v_1 u_1^\top$ to the matrix $0 \neq A \in \mathbb{R}^{n\times m}$ in the sense that

$$(\sigma_1, v_1, u_1) \in \operatorname{argmin}\left\{\frac{1}{2}\|A - \sigma v u^\top\|_2^2 \;\middle|\; \sigma \in \mathbb{R}, v \in \mathbb{S}^{n-1}, u \in \mathbb{S}^{m-1}\right\}.$$

Noticing that u and v belong to a compact set (the unit sphere), we first focus on the dependence on σ by fixing u and v. Computing the norm, we obtain

$$\begin{aligned}
\|A - \sigma v u^\top\|_2^2 &= \|A\|_2^2 - \sigma \operatorname{Tr}(u v^\top A + A^\top v u^\top) + \sigma^2 \operatorname{Tr}(u v^\top v u^\top) \\
&= \|A\|_2^2 - \sigma \sum_{i=1}^{m}\left[\sum_{j=1}^{n} u_i v_j A_i^j + \sum_{j=1}^{n} (A^\top)_j^i v_j u_i\right] + \sigma^2 \operatorname{Tr}(u u^\top) \\
&= \|A\|_2^2 - 2\sigma v^\top A u + \sigma^2.
\end{aligned}$$

This expression attains its minimum at $\sigma = v^\top A u$ so that

$$\frac{1}{2}\|A - \sigma v u^\top\|_2^2 \geqslant \frac{1}{2}[\|A\|_2^2 - (v^\top A u)^2] = \frac{1}{2}[\|A\|_2^2 - \sigma^2(u, v)] = E(u, v),$$

for any choice of $\sigma \in \mathbb{R}$, $u \in \mathbb{S}^{m-1}$, and $v \in \mathbb{S}^{n-1}$ where equality holds when σ has the special value $\sigma(u, v) = v^\top A u$. Observe that

$$\sigma(u, v) \leqslant \|A\|_2, \quad u \in \mathbb{S}^{m-1},\ v \in \mathbb{S}^{n-1},$$

thanks to the Cauchy–Schwarz inequality. By compactness of $\mathbb{S}^{m-1} \times \mathbb{S}^{n-1}$ and continuity of the map E, there is a minimum $(u_1, v_1) \in \mathbb{S}^{m-1} \times \mathbb{S}^{n-1}$ with corresponding $\sigma_1 = \sigma(u_1, v_1)$. Notice that $\sigma(u_1, v_1) = 0$ is only possible if $v^\top A u = 0$ for all u and v (why?), in which case $A = 0$. Thus unless $A = 0$, any minimizing σ does not vanish and can be assumed to be positive (otherwise replace u by $-u$ or v by $-v$).

At a minimum (u_1, v_1), we must have that all derivatives of E in directions tangential to the product manifold $\mathbb{S}^{m-1} \times \mathbb{S}^{n-1}$ vanish (remember that directional derivatives at a point are derivatives along curves on the manifold through that point). This implies that $\nabla E(u_1, v_1)$ points in a direction normal to $\mathbb{S}^{m-1} \times \mathbb{S}^{n-1}$ (i. e., orthogonal to all tangent vectors) at the point (u_1, v_1). Since the vector (u_1, v_1) is normal to $\mathbb{S}^{m-1} \times \mathbb{S}^{n-1}$ at (u_1, v_1), we see that

$$\nabla E(u_1, v_1) = \begin{bmatrix} -\sigma_1 A^\top v_1 \\ -\sigma_1 A u_1 \end{bmatrix} = \lambda \begin{bmatrix} u_1 \\ v_1 \end{bmatrix},$$

for some $0 \neq \lambda \in \mathbb{R}$, from which we infer that

$$\begin{aligned} u_1^\top(-\sigma_1 A^\top v_1) &= -\sigma_1^2 = \lambda u_1^\top u_1 = \lambda, \\ v_1^\top(-\sigma_1 A u_1) &= -\sigma_1^2 = \lambda v_1^\top v_1 = \lambda, \end{aligned}$$

whence $\lambda = -\sigma_1^2$. It follows, in particular, that

$$A^\top v_1 = \sigma_1 u_1 \quad \text{and} \quad A u_1 = \sigma_1 v_1.$$

Next, we replace A by $A_1 = A - \sigma_1 v_1 u_1^\top$ and notice that $\operatorname{rank}(A_1) < \operatorname{rank}(A)$. The latter follows from the fact that $Au_1 = \sigma_1 v_1 \neq 0$, while

$$A_1 u_1 = [A - \sigma_1 v_1 u_1^\top] u_1 = \sigma_1 v_1 - \sigma_1 v_1 = 0,$$

and the rank-nullity theorem.[1] At this point, we either have that $A_1 = 0$ in which case we are done, since $A = \sigma_1 v_1 u_1^\top$, or $A_1 \neq 0$, in which case we can repeat the argument with A replaced with A_1 and obtain another singular value $\sigma_2 > 0$ for A_1 as well as unit vectors $u_2 \in \mathbb{R}^m$ and $v_2 \in \mathbb{R}^n$ such that

$$\operatorname{rank}(A_2) < \operatorname{rank}(A_1),$$

for $A_2 = A_1 - \sigma_2 v_2 u_2^\top$. Continuing in this fashion, after $\operatorname{rank}(A)$ steps, we must have that

$$0 = A_k = A_{k-1} - \sigma_k v_k u_k^\top = A_{k-2} - \sigma_{k-1} v_{k-1} u_{k-1}^\top - \sigma_k v_k u_k^\top = \cdots$$

1 This is the statement that $\dim(\ker(A)) + \dim(\operatorname{rank}(A)) = m$ for any $A \in \mathbb{R}^{n\times m}$, which the reader is invited to prove.

$$= A - \sum_{j=1}^{k} \sigma_j v_j u_j^\top .$$

This yields the following.

Fact. *For any given matrix $A \in \mathbb{R}^{n\times m}$ of rank k, there are singular values $\sigma_1 \geqslant \sigma_2 \geqslant \cdots \geqslant \sigma_k > 0$ and singular vectors $u_1, \ldots, u_k \in \mathbb{S}^{m-1}$ and $v_1, \ldots, v_k \in \mathbb{S}^{n-1}$ such that $A = \sum_{j=1}^{k} \sigma_j v_j u_j^\top$.*

Defining matrices $U \in \mathbb{R}^{m\times k}$ by using the vectors $u_1, \ldots, u_k \in \mathbb{S}^{m-1}$ as columns and $V \in \mathbb{R}^{n\times k}$ by using the vectors $v_1, \ldots, v_n \in \mathbb{S}^{n-1}$ as columns, we see that

$$A = V\Sigma U^\top ,$$

if $\Sigma \in \mathbb{R}^{k\times k}$ denotes the diagonal matrix with the singular values as the diagonal entries. Try to show that the vectors $u_1, \ldots, u_k$ are (can be chosen to be) pairwise orthogonal as are the vectors $v_1, \ldots, v_k$. If you run into difficulties, revisit this issue after reading the next section.

8.2 Eigenvalues of symmetric matrices

In the special case of self-maps of a finite-dimensional vector space, which upon introduction of a basis, amounts to the case of square matrices, it is possible to consider eigenvalues (as opposed to the singular values of the last section). Given $A \in \mathbb{R}^{m\times m}$, this corresponds to attempting to find values $\lambda_j \in \mathbb{C}$ and vectors $u_j \in \mathbb{S}^{m-1}$ such that

$$A = \sum_{j=1}^{m} \lambda_j u_j u_j^\top .$$

Complex eigenvalues (and eigenvectors) need to be allowed as the matrix

$$A = \begin{bmatrix} 0 & -1 \\ 1 & 0 \end{bmatrix}$$

indicates. The above representation is not always possible due to a potential lack of a sufficient number of eigenvectors as the simple example

$$A = \begin{bmatrix} 0 & 1 \\ 0 & 0 \end{bmatrix}$$

shows. This matrix has indeed the single eigenvalue $\lambda = 0$ and a one-dimensional eigenspace spanned by the vector $\left[\begin{smallmatrix} 1 \\ 0 \end{smallmatrix}\right]$, which makes it impossible to obtain a basis of eigenvectors. When λ_j is real for each $j = 1, \ldots, m$, the matrix $A = \sum_{j=1}^{m} \lambda_j u_j u_j^\top$ is *symmetric* (or *self-adjoint*) in the sense that $A = A^\top$. Interestingly, the converse is also true: if a real matrix is symmetric, then it has a complete set of real eigenvalues.

Fact. *A matrix $A \in \mathbb{R}^{m\times m}$ for $m \in \mathbb{N}$ with $A = A^\top$ possesses m real eigenvalues*

$$\lambda_1 \leqslant \lambda_2 \leqslant \cdots \leqslant \lambda_m,$$

(not necessarily distinct) and an orthonormal basis of eigenvectors $u_1, \ldots, u_m$, i. e., vectors satisfying

$$Au_j = \lambda_j u_j \quad \text{and} \quad u_j^\top u_k = \delta_{jk} \quad \text{for } j,k = 1,\ldots,m,$$

such that

$$A = \sum_{j=1}^{m} \lambda_j u_j u_j^\top,$$

where it is allowed that one or more of the eigenvalues vanish.

Given the possible presence of complex eigenvalues, we need to consider the complex inner product $u^\top \overline{v}$ for $u, v \in \mathbb{C}^m$, which extends the real one on $\mathbb{R}^m \subset \mathbb{C}^m$. We then observe that

$$\overline{\lambda}|u|_2^2 = u^\top \overline{\lambda u} = u^\top \overline{Au} = \overline{(Au)^\top \overline{u}} = \overline{u^\top A^\top \overline{u}} = \overline{u}^\top \overline{A\overline{u}} = \overline{u}^\top Au = \lambda |u|_2^2,$$

since

$$u^\top \overline{v} = \overline{v^\top \overline{u}},$$

for $u, v \in \mathbb{C}^m$ and provided that u is an eigenvector of A to the eigenvalue λ, i. e., provided that $Au = \lambda u$ and $u \neq 0$. It follows that $\overline{\lambda} = \lambda$ and, therefore, that $\lambda \in \mathbb{R}$. Next, observe that

$$\lambda v^\top u = v^\top(\lambda u) = v^\top Au = u^\top A^\top v = u^\top Av = \mu u^\top v,$$

so that, if $\lambda \neq \mu$ are eigenvalues with eigenvectors u and v, respectively, it must hold that $(\lambda - \mu)v^\top u = 0$, and hence that $v^\top u = 0$, i. e., u and v must be orthogonal. If linearly independent eigenvectors can be found to the same eigenvalues, then they can be chosen to be orthonormal by applying the Gram–Schmidt orthogonalization procedure of Section 3.3.2. The existence of eigenvalues can be handled in a way similar to that used to produce singular values. Consider the optimization problem

$$\operatorname{argmin}_{\lambda\in\mathbb{R},u\in\mathbb{S}^{m-1}} \underbrace{\frac{1}{2}\|A - \lambda uu^\top\|_2^2}_{=F(\lambda,u)}.$$

A direct computation yields that

$$\frac{1}{2}\|A - \lambda uu^\top\|_2^2 = \frac{1}{2}\|A\|_2^2 - \lambda u^\top Au + \frac{1}{2}\lambda^2 \geqslant \frac{1}{2}\|A\|_2^2 - \frac{1}{2}(u^T Au)^2 = E(u),$$

since the expression is minimized at $\lambda = u^\top Au$ for any fixed u. By continuity of the map $E : \mathbb{S}^{m-1} \to \mathbb{R}$ and compactness of $\mathbb{S}^{m-1}$, a vector $u_1 \in \mathbb{S}^{m-1}$ is found that minimizes E. Then $(\lambda_1, u_1) = (u_1^\top Au_1, u_1)$ is a minimizer of F. Notice that the optimality condition amounts to the validity of

$$\nabla E(u_1) = -(u_1^\top Au_1)Au_1 = \sigma u_1$$

for some $\sigma \in \mathbb{R}$, since the gradient has to point in a direction normal to $\mathbb{S}^1$ at u_1 and the only such direction is u_1 itself (see the analogous discussion in the previous section if you need a more detailed explanation). Scalar multiplying with u_1 gives $-(u_1^\top Au_1)^2 = \sigma$, and consequently, that $Au_1 = \lambda_1 u_1$. Thus u_1 is indeed an eigenvector to the eigenvalue λ_1. By replacing $\mathbb{S}^{m-1}$ with $\mathbb{S}^{m-1} \cap V_1^\perp$ in the optimization problem, where $V_1 = \mathbb{R}u_1$, another minimizer $(\lambda_2, u_2) \in \mathbb{R} \times \mathbb{S}^{m-1}$ can be found that is orthogonal to u_1. Here, we used the term $W^\perp$ to denote the orthogonal complement subspace

$$W^\perp = \{u \in \mathbb{R}^m \mid w^\top u = 0 \;\forall w \in W\} \subset \mathbb{R}^m$$

of a subspace W of $\mathbb{R}^m$. The procedure can be repeated until $u_m \in \mathbb{S}^{m-1}$ is found, which is orthogonal to all previous minimizers $u_1, \ldots, u_{m-1}$, at which point we have a (orthonormal) basis of $\mathbb{R}^m$ and it must therefore hold that

$$A = \sum_{j=1}^{m} \lambda_j u_j u_j^\top .$$

Notice that $\infty > |\lambda_1| \geqslant |\lambda_2| \geqslant \cdots \geqslant |\lambda_m| \geqslant 0$ by construction and that some or all eigenvalues can vanish. $\surd$

If all eigenvalues of a symmetric matrix $A \in \mathbb{R}^{n \times n}$ are positive, then convince yourself that $x^\top Ax$ is a norm $|\cdot|_A$ on $\mathbb{R}^n$. The boundary of the unit ball $\mathbb{B}_{|\cdot|_A}(0, 1)$, an ellipsoid, is shown below for a particular $A \in \mathbb{R}^{2\times 2}$.

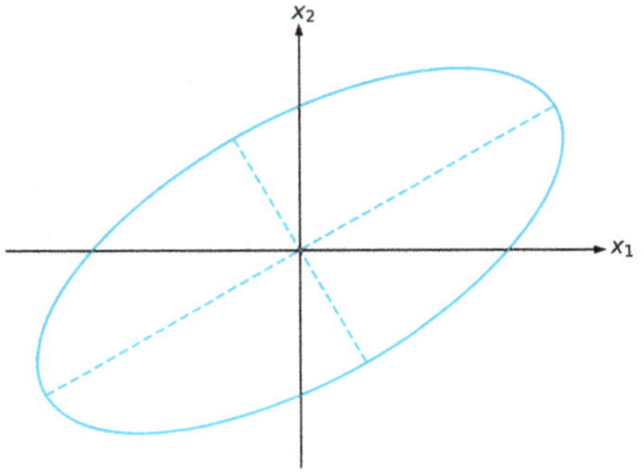

Why is it an ellipse and how do the directions and lengths of its principal axes (dashed segments) relate to the matrix A? (Q1)

Try to convince yourself that the singular values of a square matrix do not need to coincide with its eigenvalues. Eigenvectors point in directions that are left invariant by the matrix (they are only stretched or compressed), whereas singular vectors point in directions of maximal stretching. Both eigen and singular values record the amount of stretching/compression. It is a good exercise to find a connection between singular value and eigenvalue decompositions of a matrix and its representation in special bases.

We conclude this section by pointing out that the case when $A \in \mathbb{C}^{m\times m}$ is a complex matrix could also be considered. In that case, symmetry has to be replaced by the property that

$$\overline{A}^T = A, \quad \text{i.e., that} \quad \overline{a_j^k} = a_k^j, \quad j,k = 1,\dots,m.$$

Matrices satisfying this condition are called *Hermitian* or *self-adjoint*. It is good practice to show a Hermitian matrix admits a basis $v_1,\dots,v_m \in \mathbb{C}^m$ of eigenvectors to (possibly repeated) real eigenvalues $\lambda_1,\dots,\lambda_m \in \mathbb{R}$ such that

$$A = \sum_{j=1}^{m} \lambda_j v_j \overline{v_j}^T.$$

8.2.1 The conjugate gradient method

Let $A \in \mathbb{R}^{n\times n}$ be a symmetric matrix and assume that all its eigenvalues are positive.[2] Then the solution of the linear system $Ax = b$ exists for any $b \in \mathbb{R}^n$ and is unique. It is the point of minimal value of the quadratic function f defined by

$$f(x) = \frac{1}{2}x^\top Ax - x^\top b, \quad x \in \mathbb{R}^n.$$

This can be verified by computing its gradient

$$\nabla f(x) = Ax - b, \quad x \in \mathbb{R}^n,$$

and noticing that its vanishing amounts to the validity of $Ax = b$. The function f is strictly convex and satisfies $\lim_{|x|_2\to\infty} f(x) = \infty$ (check these statements[3]). In the next section, we shall see that this implies the existence of a unique point of minimum. In applications, one is often confronted with the problem of inverting large matrices. Direct inversion via some formula or Gauss elimination can prove computationally

2 One says that A is positive definite when this holds and writes $A = A^\top > 0$.

3 If you are not familiar with convexity, take a peek at the beginning of the next section.

intensive and iterative methods can provide a viable alternative. The simplest iteration for the above problem is obtained by applying gradient descent to f. Starting with an initial guess $x_0 \in \mathbb{R}^n$, one iterates based on

$$x_{k+1} = x_k - \alpha \nabla f(x_k) = x_k + \alpha(b - Ax_k) =: x_k + \alpha r_k,$$

where r_k is called the *residual*. The step α is chosen to be optimal, i. e., to satisfy

$$\begin{aligned} 0 &= \frac{d}{d\alpha} f(x_k + \alpha r_k) = \nabla f(x_k + \alpha r_k) \cdot r_k \\ &= (Ax_k + \alpha A r_k - b) \cdot r_k = \alpha r_k^\top A r_k - r_k^\top r_k. \end{aligned}$$

This yields the algorithm:

1. Choose x_0.
2. Repeat for k=0,1,2,...

$$r_k = b - Ax_k, \quad \alpha_k = \frac{r_k^\top r_k}{r_k^\top A r_k}$$

$$x_{k+1} = x_k + \alpha_k r_k$$

until r_k is as small as desired.

Observe that $\nabla f(x_{k+1}) \cdot r_k = 0$ and, more in general, that

$$\nabla f(x + v_*) \cdot v = 0 \quad \text{for } v \in U,$$

if $U \subset \mathbb{R}^n$ is any subspace and if $v_* \in \operatorname{argmin}_{v \in U} f(x+v)$. Given $v \in U$, this follows from

$$g'(0) = 0 \quad \text{for } g(t) = f(x + v_* + tv), \ t \in \mathbb{R}.$$

This method can be slow as it may bounce off steep walls as it slowly makes its way down a narrow canyon. This happens when the matrix A has eigenvalues of significant differing sizes so that the ellipsoidal level sets of f are elongated. In the two-dimensional depiction below, the path taken by gradient descent appears to zig-zag about the level lines of f toward the solution (i. e., the minimizer) given by the black dot.

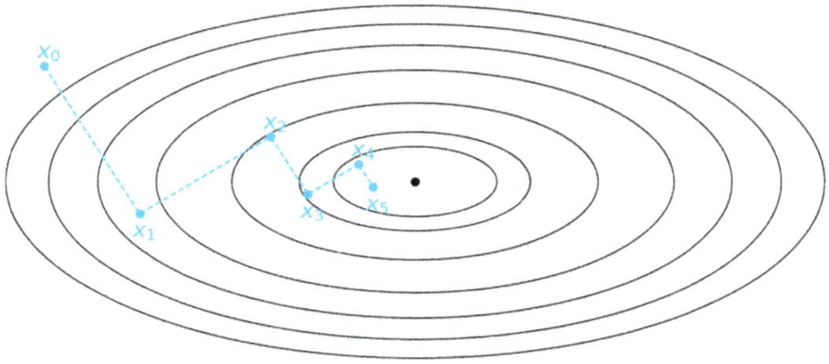

It is possible to improve this algorithm by sequentially minimizing in conjugate directions, i. e., in directions which are A-orthogonal[4] as opposed to orthogonal with respect to the standard Euclidean inner product. We start, as in plain gradient descent, with an initial guess x_0 and find the point of minimum x_1 of f along the line through x_0 with direction given by the (negative) gradient $-\nabla f(x_0) = b - Ax_0$. It is given by

$$x_1 = x_0 + \alpha_*^0 r_0 \quad \text{for } \alpha_*^0 = \frac{r_0^\top r_0}{r_0^\top A r_0},$$

and, as we only have one direction r_0 for now, set $p_0 = r_0$ for the first vector in the A-orthogonal basis. Next, observe that, for any basis consisting of A-orthogonal vectors $p_0, \dots, p_{n-1}$, it holds that

$$\operatorname*{argmin}_{\alpha^0,\dots,\alpha^{n-1}} f\left(x_0 + \sum_{j=0}^{n-1} \alpha^j p_j\right) = \left(\alpha_*^0, \operatorname*{argmin}_{\alpha^1,\dots,\alpha^{n-1}} f\left(x_1 + \sum_{j=1}^{n-1} \alpha^j p_j\right)\right),$$

since

$$\begin{aligned} 0 &= \partial_{\alpha^0} f\left(x_0 + \sum_{j=0}^{n-1} \alpha^j p_j\right) = \nabla f\left(x_0 + \sum_{j=0}^{n-1} \alpha^j p_j\right) \cdot p_0 \\ &= \left(Ax_0 - b + \sum_{j=0}^{n-1} \alpha^j A p_j\right) \cdot p_0 = (-p_0 + \alpha^0 A p_0) \cdot p_0 \\ &= \partial_{\alpha^0} f(x_0 + \alpha^0 p_0), \end{aligned}$$

where the fourth identity follows from $p_j^\top A p_0 = 0$ for $j = 1, \dots, n-1$. This means that minimizing f along the line $\{x_0 + \alpha^0 r_0 \mid \alpha^0 \in \mathbb{R}\}$ yields the same minimizing value α_*^0 as we would get by minimizing $f(x_0 + \sum_{j=0}^{n-1} \alpha^j p_j)$ over all $\alpha^0, \dots, \alpha^{n-1}$ and extracting the first component of the minimizer. We can now continue with $r_1 = b - Ax_1 = -\nabla f(x_1)$ and modify it to p_1 making it A-orthogonal to r_0 by a Gram–Schmidt orthogonalization step, i. e.,

$$p_1 = r_1 - \frac{r_1^\top A p_0}{p_0^\top A p_0} p_0.$$

Minimizing $f(x_1 + \alpha^1 p_1)$ in α^1 yields

$$x_2 = x_1 + \alpha_*^1 p_1 \quad \text{with } \alpha_*^1 = \frac{p_1^\top p_1}{p_1^\top A p_1}.$$

4 Two vectors x, y are A-orthogonal if $x^\top A y = 0$, i. e., orthogonal with respect to the scalar product induced by A.

Notice that $r_0 \in \operatorname{span}\{r_0\}$ and that

$$r_1 = b - Ax_1 = b - Ax_0 - \alpha_*^0 Ar_0 = r_0 - \alpha_*^0 Ar_0 \in \operatorname{span}\{r_0, Ar_0\},$$

as well as that

$$\operatorname{span}\{r_0, r_1\} = \operatorname{span}\{r_0, Ar_0\} = \operatorname{span}\{p_0, p_1\}.$$

Using this and the fact that

$$(\alpha_*^0, \alpha_*^1) = \operatorname*{argmin}_{\alpha^0,\alpha^1} f(x_0 + \alpha^0 p_0 + \alpha^1 p_1),$$

we see that

$$r_2 = -\nabla f(x_2) \perp \operatorname{span}\{r_0, Ar_0\}.$$

In particular, $r_2^\top Ap_0 = 0$ and, when modifying $r_2 = b - Ax_2$ to obtain p_2, which is A-orthogonal to p_0 and p_1, it suffices to obtain A-orthogonality to p_1. This gives

$$p_2 = r_2 - \frac{r_2^\top Ap_1}{p_1^\top Ap_1} p_1,$$

and the procedure can continue. Depending on the initial guess x_0, the process will end when $r_i = 0$, which can happen for any $i \in \{0, \dots, n-1\}$. In any event, after at most n steps, the minimizer is identified. We obtained the following *conjugate gradient algorithm:*

1. Choose x_0 and set $p_{-1} = 0$.
2. Repeat for $k = 0, 1, 2, \dots, n-1$

$$r_k = b - Ax_k, \quad p_k = r_k - \frac{r_k^\top Ap_{k-1}}{p_{k-1}^\top Ap_{k-1}} p_{k-1}, \quad \alpha_k = \frac{r_k^\top p_k}{p_k^\top Ap_k}$$

$$x_{k+1} = x_k + \alpha_k p_k.$$

In the two-dimensional example used to illustrate gradient descent, the conjugate gradient algorithm obtains the solution in two steps as depicted below

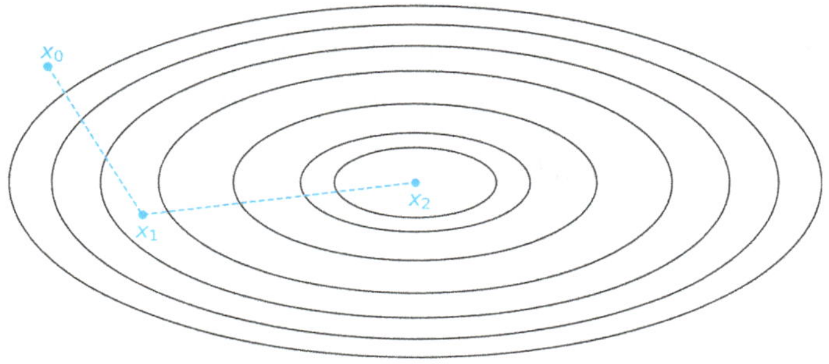

In the derivation of the algorithm, we made use of the fact that r_k is A-orthogonal to $p_0, \dots, p_{k-2}$ for $k = 2, 3, \dots, n-1$. A proof was given for $k = 2$ in the argument above. Try to obtain one for $k > 2$.[5]

8.3 Convexity

Convexity is a structure that, when available, opens the door to a whole set of tools that prove very useful in many areas of mathematics where optimization plays a role. Its tools are known under the umbrella of *convex analysis*. Here, we simply would like to introduce the concept and offer a peek into some basic but important consequences it entails. Given a subset U of a vector space V, we call it *convex* iff

$$\begin{aligned} U \supset [x, y] &= \{x + t(y - x) \mid t \in [0, 1]\} \\ &= \{(1 - t)x + ty \mid t \in [0, 1]\} \quad \forall x, y \in U. \end{aligned}$$

This simply means that a set is convex if the segment $[x, y]$ connecting any two of its points x, y is contained in the set itself. Vector (sub)spaces are clearly convex, since they contain the whole line determined by at any two of their points. In $\mathbb{R}^d$ for $d \in \mathbb{N}$, the balls

$$\mathbb{B}_p(0, 1) = \{x \in \mathbb{R}^d \mid |x|_p < 1\}$$

with respect to the p-norm given by $|x|_p = (\sum_{k=1}^d |x_k|^p)^{1/p}$ for $p \in [1, \infty)$ and given by $|x|_\infty = \max_{k=1,\dots,d} |x_k|$ for $p = \infty$ and $x \in \mathbb{R}^d$, are convex (check this). An annulus like

$$\mathbb{B}_2(0, 2) \setminus \mathbb{B}_2(0, 1) \subset \mathbb{R}^d$$

is an example of a nonconvex set. If $L \in \mathcal{L}(V, W)$ is a linear map between vector spaces and $U \subset V$ is convex, so is $L(U) \subset W$. Indeed, given vectors $w_1, w_2 \in L(U)$, there are $v_1, v_2 \in U$ with $L(v_j) = w_j$ for $j = 1, 2$. Then, for $t \in [0, 1]$,

$$(1 - t)w_1 + tw_2 = (1 - t)L(v_1) + tL(v_2) = L((1 - t)v_1 + tv_2) \in L(U),$$

since $(1 - t)v_1 + tv_2 \in U$ by the convexity of U. For the rest of this section, we assume that $V = \mathbb{R}^d$ since the infinite-dimensional case will not be discussed. The set

$$\operatorname{epi}(f) = \{(x, t) \in \mathbb{R}^d \times \mathbb{R} \mid t \geq f(x)\}$$

5 It would be natural to try and do so by induction.

is called the *epigraph* of the function $f : \mathbb{R}^d \to \overline{\mathbb{R}}$, where $\overline{\mathbb{R}} = \mathbb{R} \cup \{\infty\}$. An example of epigraph is depicted in the image below.

The epigraph of a function f with $f(x) = \infty$ on $[-1, 1]^c$

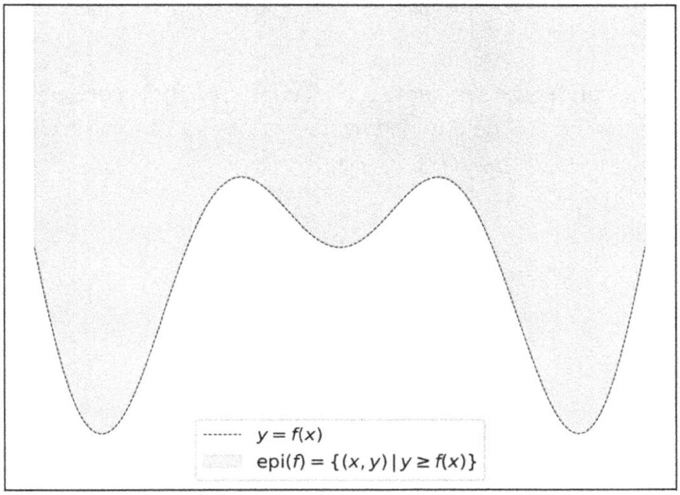

A function $f : \mathbb{R}^d \to \overline{\mathbb{R}}$ is called *convex* iff its epigraph is convex. Given such a convex function, the set

$$\mathrm{dom}(f) = \{f < \infty\} = \{x \in \mathbb{R}^d \mid f(x) < \infty\}$$

is called the *effective domain* of f. If $x, y \in \mathrm{dom}(f)$ for a convex function f, then $(x, f(x)), (y, f(y)) \in \mathrm{epi}(f)$, so that

$$((1-t)x + ty, (1-t)f(x) + tf(y)) \in \mathrm{epi}(f) \quad \forall t \in [0, 1],$$

or, equivalently, that

$$f((1-t)x + ty) \leqslant (1-t)f(x) + tf(y) < \infty \quad \forall t \in [0, 1],$$

showing that $\mathrm{dom}(f)$ is convex for any convex function f. This can also be seen by observing that $\mathrm{dom}(f) = P_1(\mathrm{epi}(f))$ for the linear map (projection) $P_1 : \mathbb{R}^d \times \mathbb{R} \to \mathbb{R}^d$, $(x, t) \mapsto x$. Given a function $f : U \to \mathbb{R}$, defined on a convex set U, it can be identified with its extension

$$\bar{f} : \mathbb{R}^d \to \overline{\mathbb{R}}, \quad x \mapsto \begin{cases} f(x), & x \in U, \\ \infty, & x \notin U. \end{cases}$$

It is called convex if $\bar{f}$ is convex, in which case $U = \mathrm{dom}(\bar{f})$. It follows from the above discussion that in this case f is convex iff it holds that

$$f((1-t)x+ty) \leqslant (1-t)f(x)+tf(y) \quad \forall t \in [0,1],$$

for $x, y \in U$.[6] The geometric interpretation of convexity is that the curve

$$\gamma^f_{x,y} = \{(z, f(z)) \mid z \in [x,y]\} \subset U \times \mathbb{R} \subset V \times \mathbb{R}$$

on the graph of f always lies under the segment $s^f_{x,y} = [(x, f(x)), (y, f(y))]$ connecting its end points in the plane $\mathbb{R}(y-x) \times \mathbb{R}$ determined by the vector $y-x$ and the vertical axis of the graph.

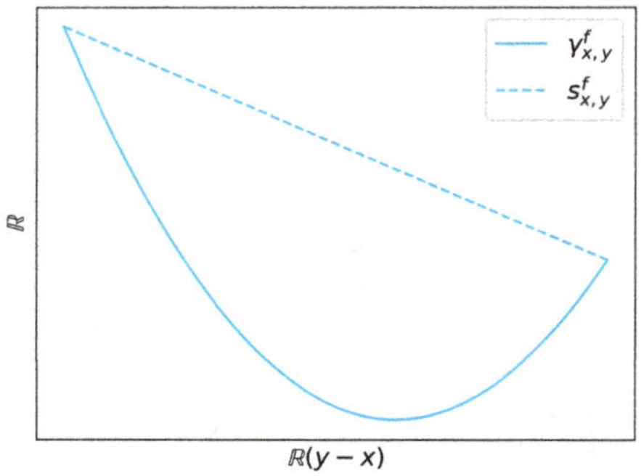

Verify that convexity of a function $f : U \to \mathbb{R}$ defined on a convex set $U \subset \mathbb{R}^d$ is equivalent to the validity of

$$f(\lambda_1 x_1 + \cdots + \lambda_n x_n) \leqslant \lambda_1 f(x_1) + \cdots + \lambda_n f(x_n),$$

for $n \in \mathbb{N}$, $x_1, \ldots, x_n \in U$, and $\lambda_k \in [0,1]$ with $\sum_{k=1}^d \lambda_k = 1$. The functions

$$|\cdot|_p^p : \mathbb{R}^d \to [0,\infty), \quad x \mapsto |x|_p^p,$$

are convex for $p \in [1,\infty)$. So are the functions

$$\exp : \mathbb{R} \to \mathbb{R},\ x \mapsto e^x \quad \text{and} \quad -\log : (0,\infty) \to \mathbb{R},\ x \mapsto -\log(x).$$

Show that convexity of $-\log$ implies the following inequality of arithmetic and geometric means:

$$(x_1 \cdots x_n)^{1/n} \leqslant \frac{x_1 + \cdots + x_n}{n},$$

valid for $x_1, \ldots, x_n \in (0,\infty)$.

6 One speaks of *strict* convexity if the defining inequality is strict.

A convex function $f : \mathbb{R}^d \to \overline{\mathbb{R}}$ has convex domain $\mathrm{dom}(f)$ that may not be full-dimensional. Take for instance $f : \mathbb{R}^2 \to \overline{\mathbb{R}}$ defined by

$$f(x) = \begin{cases} 0, & \text{if } x_1 \in (0,1) \text{ and } x_2 = 0, \\ \infty, & \text{otherwise,} \end{cases}$$

for which $\mathrm{dom}(f) = (0,1)\times\{0\}$. In such a situation, we can simply "discard" the dimensions in which the function is identically equal to infinity and assume without loss of generality that the domain is full-dimensional. If $\mathrm{dom}(f) \neq \emptyset$, i. e., if $f \not\equiv \infty$, and there are at least two points in $\mathrm{dom}(f)$, then the domain has nonempty interior $\mathrm{int}(\mathrm{dom}(f))$. Now consider

$$f(x + th) = f((1-t)x + t(x+h)) \leqslant (1-t)f(x) + tf(x+h),$$

for $x \in \mathrm{int}(\mathrm{dom}(f))$, $h \in \mathbb{R}^d$, and t small enough so that $x + th \in \mathrm{int}(\mathrm{dom}(f))$. The inequality rewrites as

$$\frac{f(x+th) - f(x)}{t} \leqslant f(x+h) - f(x), \tag{8.1}$$

and implies that

$$\partial_h f(x) = \nabla f(x) \cdot h \leqslant f(x+h) - f(x),$$

upon letting t tend to 0, provided f is differentiable at x. This shows that the graph of a differentiable convex function f defined on an open set U, always lies above its tangent (hyper)plane, i. e., that

$$f(x+h) \geqslant f(x) + \nabla f(x) \cdot h, \quad h \in \mathbb{R}^d.$$

If f is not differentiable at a point x with $f(x) < \infty$, it still holds that

$$f(x + \tau h) = f((1-\tau)x + \tau(x+h)) \leqslant (1-\tau)f(x) + \tau f(x+h),$$

for $\tau \in (0,1]$, which can be rewritten as

$$\frac{f(x+\tau th) - f(x)}{\tau t} \leqslant \frac{f(x+th) - f(x)}{t},$$

where $t \in (0,\infty)$ and $\tau \in (0,1]$ by replacing h with th. This shows that the map

$$d_h f(x,\cdot) : (0,\infty) \to \mathbb{R}, \quad t \mapsto \frac{f(x+th) - f(x)}{t}$$

is monotone nondecreasing, so that

$$d_h f(x,0) = \lim_{t \searrow 0} d_h f(x,t) \in \overline{\mathbb{R}},$$

is well-defined for any $h \in \mathbb{R}^d$ and is the directional derivative of f at x in direction h. Show that $d_{sh} f(x,0) = s\, d_h f(x,0)$ for $s > 0$, that $d_h f(x,0)$ is a convex function of h, and consequently, that

$$-d_{-h} f(x,0) \leqslant d_h f(x,0) \quad \text{for } h \in \mathbb{R}^d.$$

If f is differentiable at x, then $d_h f(x,0) = \partial_h f(x) = \nabla f(x) \cdot h$. A vector $v \in \mathbb{R}^d$ is called a *subgradient* of f at x iff it holds that

$$f(x+h) \geqslant f(x) + v \cdot h \quad \forall h \in \mathbb{R}^d,$$

and the set $\partial f(x)$ of all subgradients of f at a point $x \in \operatorname{dom}(f)$ is called *subdifferential*. A function is called *subdifferentiable* at $x \in \operatorname{dom}(f)$ iff $\partial f(x) \neq \emptyset$. When a function f is differentiable at $x \in \operatorname{dom}(f)$, then $\partial f(x) = \{\nabla f(x)\}$.

The simplest but still very illustrative example is given by the absolute value function $|\cdot|$, which is subdifferentiable everywhere and differentiable for $x \neq 0$. At $x = 0$, we have that

$$\partial|\cdot|(0) = [-1, 1],$$

which is compatible with the observation that any line through the origin with slope between -1 and 1 is "tangent" to the graph of $|\cdot|$.

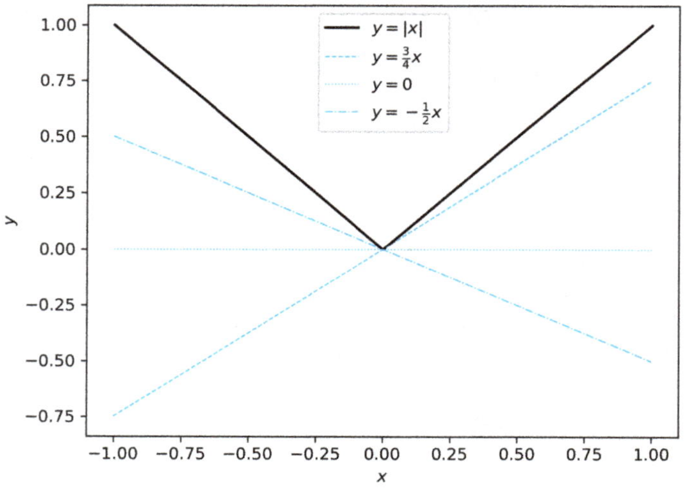

Notice how the extreme values of $\partial|\cdot|(0)$ correspond to the directional derivatives of the function at $x = 0$. Compute the subdifferential of the function

$$|\cdot|_1 : \mathbb{R}^d \to \mathbb{R}, \quad x \mapsto |x|_1 = \sum_{k=1}^{d} |x_k|,$$

discuss (Q2) the correspondence with the absolute value ($d = 1$), and the relation between $\partial|\cdot|_1(0)$ and $\partial|\cdot|_1(x) = \{(\operatorname{sign}(x_1), \dots, \operatorname{sign}(x_d))\}$ for x such that $\prod_{j=1}^{d} x_j \neq 0$. What is the subdifferential of $|\cdot|_\infty$ given by

$$|x|_\infty = \max_{k=1,\dots,d} |x_k|, \quad x \in \mathbb{R}^d \text{ for } d > 1?$$

The function $f : \mathbb{R} \to \mathbb{R}$ defined by

$$f(x) = \begin{cases} -\sqrt{1-x^2} & \text{for } |x| \leqslant 1, \\ \infty & \text{for } |x| > 1, \end{cases}$$

shows that a convex function can fail to be subdifferentiable at points in its domain (here for $x = \pm 1$). Given a convex function $f : \mathbb{R}^d \to \overline{\mathbb{R}}$ and $x \in \operatorname{dom}(f)$, show that $\mathbb{R}^d \ni v \in \partial f(x)$ iff

$$v \cdot h \leqslant d_h f(x) \quad \forall h \in \mathbb{R}^d.$$

A remarkable property of convex functions (defined on a finite-dimensional space) is that they are automatically continuous.

Fact. *Let $f : \mathbb{R}^d \to \overline{\mathbb{R}}$ be convex and assume that* $\operatorname{dom}(f) \neq \emptyset$ *is open. Then the function $f : \operatorname{dom}(f) \to \mathbb{R}$ is continuous. It is actually locally Lipschitz continuous.*[7]

We will need inequality (8.1), which reads

$$f(x + th) - f(x) \leqslant t[f(x + h) - f(x)],$$

and the similar

$$t[f(x) - f(x - h)] \leqslant f(x + th) - f(x),$$

which hold for any $x \in \operatorname{dom}(f)$, $h \in \mathbb{R}^d$ with $x \pm h \in \operatorname{dom}(f)$, and $t \in [0, 1]$. The second inequality, just as the first, is a consequence of convexity if we observe that

$$x = \frac{t}{1+t}(x - h) + \frac{1}{1+t}(x + th).$$

7 Local Lipschitz continuity amounts to Lipschitz continuity in the neighborhood of each point.

Indeed

$$f(x) \leqslant \frac{t}{1+t} f(x-h) + \frac{1}{1+t} f(x+th)$$

is equivalent to it. Given $x \in \operatorname{dom}(f)$ it is possible to find $t_x \in (0,1)$ such that

$$x + \overline{\mathbb{B}}_{|\cdot|_1}(0,t_x) = x + \left\{ y \in \mathbb{R}^d \;\middle|\; |y|_1 = \sum_{k=1}^{d} |y^k| \leqslant t_x \right\} \subset \operatorname{dom}(f)$$

since the effective domain is assumed to be open. Any point in $x + \overline{\mathbb{B}}_{|\cdot|_1}(0,t_x)$ is of the form

$$x + th \text{ for } t \leqslant t_x \text{ and } h \text{ with } |h|_1 = 1.$$

It follows that

$$t\big[f(x) - f(x-h)\big] \leqslant f(x+th) - f(x) \leqslant t\big[f(x+h) - f(x)\big].$$

Using

$$x \pm h = x \pm \sum_{k=1}^{d} \underbrace{\operatorname{sign}(h^k)}_{s^k} |h^k| e_k = \sum_{k=1}^{d} |h^k| [x \pm s^k e_k],$$

and convexity, we arrive at

$$\begin{aligned} f(x+h) - f(x) &\leqslant \sum_{k=1}^{d} |h^k| [f(x + s^k e_k) - f(x)] \\ &\leqslant \max_{k=1,\dots,d;\; s=\pm 1} f(x + s e_k) - f(x) = M \end{aligned}$$

and at

$$\begin{aligned} f(x) - f(x-h) &\geqslant f(x) - \sum_{k=1}^{d} |h^k| f(x - s^k e_k) \\ &\geqslant f(x) - \max_{k=1,\dots,d;\; s=\pm 1} f(x - s e_k) = -M \end{aligned}$$

since $|h|_1 = \sum_{k=1}^{d} |h^k| = 1$. This, together with the inequalities for $f(x+th) - f(x)$ derived above, implies

$$\big|f(x+th) - f(x)\big| \leqslant Mt \quad \text{for } t \leqslant t_x \text{ and } |h|_1 = 1,$$

and continuity follows. This is, however, not local Lipschitz continuity yet. To obtain the latter, fix $x \in \operatorname{dom}(f)$ and $t > 0$ with $\overline{\mathbb{B}}_{|\cdot|_2}(x,t) \subset \operatorname{dom}(f)$, and notice that

$$|f(w)| \leqslant M \quad \text{for } w \in \overline{B}_{|\cdot|_2}(x,t),$$

for some $M > 0$ by continuity of f and compactness of the closed ball. For $\varepsilon \in (0,t)$, take $y,z \in \overline{\mathbb{B}}_{|\cdot|_2}(x,\varepsilon)$ and define

$$w = y + \frac{t-\varepsilon}{|y-z|_2}(y-z) \in \overline{\mathbb{B}}_{|\cdot|_2}(x,t).$$

Algebraic manipulations show that

$$y = \frac{t-\varepsilon}{|y-z|_2 + t - \varepsilon} z + \frac{|y-z|_2}{|y-z|_2 + t - \varepsilon} w,$$

and convexity yields

$$f(y) \leqslant \frac{(t-\varepsilon)}{|y-z|_2 + t - \varepsilon} f(z) + \frac{|y-z|_2}{|y-z|_2 + t - \varepsilon} f(w),$$

which rewrites as

$$(t-\varepsilon)[f(y) - f(z)] \leqslant [f(w) - f(y)]|y-z|_2 \leqslant 2M|y-z|_2.$$

By switching the role of y and z, we arrive at

$$|f(y) - f(z)| \leqslant L|y-z|_2 \quad \forall y,z \in \overline{\mathbb{B}}_{|\cdot|_2}(0,\varepsilon),$$

with $L = 2M/(t-\varepsilon)$, i. e., at (local) Lipschitz continuity about the point $x \in \operatorname{dom}(f)$, which was arbitrary. $\surd$

The convex function $f : \mathbb{R} \to \overline{\mathbb{R}}$ given by

$$f(x) = \begin{cases} 0, & x \in (-1,1), \\ 1, & x = \pm 1, \\ \infty, & x \in [-1,1]^c, \end{cases}$$

shows that continuity may fail on the boundary of the effective domain (and is the reason we assume that it be open for the continuity result). In order to exclude this possibility a convex function with (full-dimensional) $\operatorname{dom}(f) \neq \emptyset$ can be modified to its so-called closure $\operatorname{cl}(f)$ by setting

$$\operatorname{cl}(f)(x) = \liminf_{y \to x} f(y) = \lim_{\delta \to 0} \inf_{|y-x|_2 \leqslant \delta} f(y),$$

which only alters f on $\partial \operatorname{dom}(f)$, given its continuity in the interior of its domain. In the concrete example above, the values of f at $x = \pm 1$ would be reset to 0 when taking the closure.

We conclude this section with the following important fact.

Fact. *Let $f : \mathbb{R}^d \to \mathbb{R}$ be strictly convex and assume that*

$$\lim_{|x|_2 \to \infty} f(x) = \infty.$$

Then f attains its unique minimum.

The assumption about the behavior of f for large arguments yields the existence of $R > 0$ such that

$$f(x) \geqslant 2f(0) \quad \text{for } |x|_2 \geqslant R.$$

We infer that the points where f has its infimum $m \in \mathbb{R}$ have to belong to $\overline{\mathbb{B}}_2(0, R)$. As we just learned that convex functions are continuous, we conclude that $f|_{\overline{\mathbb{B}}_2(0,R)}$ attains its minimum in at least one point. Finally, assuming that the minimum is taken on at two points $x_1, x_2 \in \overline{\mathbb{B}}_2(0, R)$ at least, we can use strict convexity to obtain

$$f\left(\frac{1}{2}x_1 + \frac{1}{2}x_2\right) < \frac{1}{2}f(x_1) + \frac{1}{2}f(x_2) = m,$$

which contradicts the fact that m is the infimum value of f. Hence the minimum is unique. The function

$$f : \mathbb{R} \to \mathbb{R}, \quad x \mapsto e^x$$

shows that the condition about the behavior at infinity is necessary since it is strictly convex but does not attain its minimum value 0.

9 Partial differential equations

In this chapter, we give a very limited and partial introduction to the treatment of the heat equation as an example of an approach that can be taken to deal with evolutionary (time dependent) equations and as motivation for the introduction of function spaces of general interest. A possible and interesting derivation of the equation is offered in the next chapter. The approach is very much inspired by that taken to solve ODEs.

9.1 The periodic heat equation

The (linear) *heat (or diffusion) equation* for a periodic function

$$u : \mathbb{T}^d = [0, 2\pi]^d \to \mathbb{C}$$

in dimension $d \in \mathbb{N}$ with periodic initial datum u_0 reads

$$\begin{cases} \partial_t u - \Delta u = 0 & \text{in } \mathbb{T}^d \text{ for } t > 0, \\ u(0, \cdot) = u_0 & \text{in } \mathbb{T}^d, \end{cases}$$

where $\Delta u = \sum_{j=1}^d \partial_j^2 u$ and periodicity means

$$u(x + 2\pi e_j) = u(x) \quad \forall x \in \mathbb{T}^d \ \forall j = 1, \dots, d.$$

We can think of this equation as the linear ODE $\dot{u} = Au$ where the matrix A (see Chapter 7) is replaced by Δ. Formally, the solution should be given by $e^{t\Delta}u_0$ and we know that a way to compute the exponential of a matrix is through the determination of eigenvalues and eigenvectors. We therefore consider the eigenvalue equation

$$\Delta u = \mu u,$$

for periodic u. Looking for periodic solutions of the form $u(x) = u(x_j)$ for $j \in \{1, \dots, d\}$, the equation reduces to $u'' = \mu u$. The general solution of the latter equation is given by $\alpha e^{\sqrt{\mu}\xi} + \beta e^{-\sqrt{\mu}\xi}$ for $\alpha, \beta \in \mathbb{C}$. Such solution is only 2π-periodic if we have that $\sqrt{\mu} = ik$ for some $k \in \mathbb{Z}$, in which case $\mu = -k^2$. We therefore have the eigenvalues $\mu_k = -k^2$, $k \in \mathbb{N}$, with associated eigenvectors (better eigenfunctions)

$$e^{\pm ik\xi}, \quad \xi \in \mathbb{T}.$$

https://doi.org/10.1515/9783110780925-009

Notice that

$$\int_0^{2\pi} e^{ik\xi}e^{-il\xi}\,d\xi = 2\pi\delta_{kl}, \quad k,l \in \mathbb{Z},$$

making

$$\left\{\varphi_k = \frac{1}{\sqrt{2\pi}}e^{ik\cdot} \mid k \in \mathbb{Z}\right\},$$

an orthonormal system on

$$C_\pi^\infty = \{u \in C^\infty(\mathbb{T},\mathbb{C}) \mid u \text{ is } 2\pi\text{-periodic}\},$$

with respect to the scalar product

$$(u|v)_2 = \int_0^{2\pi} u(\xi)\overline{v}(\xi)\,d\xi, \quad u,v \in C_\pi^\infty.$$

Choosing $k \in \mathbb{Z}^d$ and defining

$$\varphi_k(x) = \frac{1}{(2\pi)^{d/2}}\prod_{j=1}^d \varphi_{k_j}(x_j),$$

yields (orthonormal) eigenfunctions of Δ as follows from

$$\Delta\varphi_k(x) = -\left[\sum_{j=1}^d k_j^2\right]\prod_{j=1}^d \varphi_{k_j}(x_j) = -|k|_2^2\varphi_k(x).$$

Inspired by the finite-dimensional case of matrices, we consider linear combinations of eigenfunctions. Since there are infinitely many of the latter, convergence considerations will have to play a role. Here, we consider the natural choice

$$\ell_2(\mathbb{Z}^d,\mathbb{C}) = \left\{(\hat{u}_k)_{k\in\mathbb{Z}^d} \;\middle|\; \hat{u}_k \in \mathbb{C} \text{ for } k \in \mathbb{Z}^d \text{ and } \sum_{k\in\mathbb{Z}^d}|\hat{u}_k|^2 < \infty\right\},$$

where absolute convergence makes it possible to add the terms in the series in any order. This vector space carries the inner product given by

$$(\hat{u} \mid \hat{v}) = \sum_{k\in\mathbb{Z}^d}\hat{u}_k\overline{\hat{v}}_k, \quad \hat{u},\hat{v} \in \ell_2(\mathbb{Z}^d,\mathbb{C}),$$

for $\hat{u} = (\hat{u}_k)_{k\in\mathbb{Z}^d}$ and $\hat{v} = (\hat{v}_k)_{k\in\mathbb{Z}^d}$.

Fact. *The space $\ell_2(\mathbb{Z}^d, \mathbb{C})$ is complete, i. e., it is a Hilbert space.*

In order to see this, take a Cauchy sequence $(x^n)_{n\in\mathbb{N}}$ in $\ell^2(\mathbb{Z}^d, \mathbb{C})$ with respect to the norm induced by the scalar product. Since for any given $k \in \mathbb{Z}^d$,

$$|x_k^n - x_k^m| \leqslant \|x^n - x^m\|_2 = \sqrt{\sum_{k\in\mathbb{Z}^d} |x_k^n - x_k^m|^2} \longrightarrow 0 \quad \text{as } m, n \to \infty,$$

the sequence $(x_k^n)_{n\in\mathbb{N}}$ in $\mathbb{C}$ of the kth component x_k^n of x^n is Cauchy and, therefore, admits a limit $x_k^\infty \in \mathbb{C}$ for each $k \in \mathbb{Z}^d$. Cauchy sequences are bounded sequences, and thus it holds that

$$\sum_{|k|_2\leqslant M} |x_k^n|^2 \leqslant \sum_{k\in\mathbb{Z}^d} |x_k^n|^2 = \|x^n\|_2^2 \leqslant C, \quad n \in \mathbb{N}$$

for some $C > 0$. Taking the limit as $n \to \infty$, this yields that

$$\lim_{n\to\infty} \sum_{|k|_2\leqslant M} |x_k^n|^2 = \sum_{|k|_2\leqslant M} |x_k^\infty|^2 \leqslant C, \quad M \in \mathbb{N},$$

and letting M tend to ∞, we see that $x^\infty \in \ell_2(\mathbb{Z}^d, \mathbb{C})$. Finally, taking the limit $m \to \infty$ in the inequality

$$\sqrt{\sum_{|k|_2\leqslant M} |x_k^n - x_k^m|^2} \leqslant \|x^n - x^m\|_2 \leqslant \varepsilon,$$

which, for any given $\varepsilon > 0$, is valid for $m, n \geqslant N_\varepsilon$, some $N_\varepsilon \in \mathbb{N}$, and arbitrary $M \in \mathbb{N}$, we arrive at

$$\sqrt{\sum_{|k|_2\leqslant M} |x_k^n - x_k^\infty|^2} \leqslant \varepsilon \quad \text{for } n \geqslant N_\varepsilon,\ M \in \mathbb{N},$$

and eventually to $\|x^n - x^\infty\|_2 \leqslant \varepsilon$ for $n \geqslant N_\varepsilon$, by letting M tend to ∞. We conclude that $x^n \to x^\infty$ in $\ell_2(\mathbb{Z}^d, \mathbb{C})$. $\checkmark$

Taking the subspace of finite sequences,

$$c_{00}(\mathbb{Z}^d, \mathbb{C}) = \{(\hat{u}_k)_{k\in\mathbb{Z}^d} \mid \hat{u}_k = 0 \text{ for all but finitely many } k \in \mathbb{Z}^d\},$$

we see that $u = \sum_{k\in\mathbb{Z}^d} \hat{u}_k \varphi_k$ for $(\hat{u}_k)_{k\in\mathbb{N}} \in c_{00}(\mathbb{Z}^d, \mathbb{C})$ satisfies that

$$u \in C_\pi^\infty \quad \text{and that} \quad \int_{\mathbb{T}^d} |u(x)|^2\, dx = \sum_{k\in\mathbb{Z}^d} |\hat{u}_k|^2,$$

due to orthonormality of $(\varphi_k)_{k\in\mathbb{N}}$. This gives a map

$$\tilde{\mathcal{G}} : c_{00}(\mathbb{Z}^d, \mathbb{C}) \to \mathrm{L}^2(\mathbb{T}^d, \mathbb{C}), \quad \hat{u} = (\hat{u}_k)_{k\in\mathbb{N}} \mapsto u = \sum_{k\in\mathbb{Z}^d} \hat{u}_k \varphi_k,$$

satisfying $\tilde{\mathcal{G}}(c_{00}(\mathbb{Z}^d, \mathbb{C})) \subset \mathrm{C}_\pi^\infty$ and $\|u\|_2 = \|u\|_{\mathrm{L}^2} = \|\hat{u}\|_2 = \|\hat{u}\|_{\ell_2}$. It can be shown (do it! (Q1)) that $\ell_2(\mathbb{Z}^d, \mathbb{C})$ is the completion of $c_{00}(\mathbb{Z}^d, \mathbb{C})$ with respect to the $\|\cdot\|_{\ell_2}$ norm. Recalling that $\mathrm{L}^2(\mathbb{T}^d)$ was defined as the completion of $\mathrm{C}(\mathbb{T}^d, \mathbb{C}) \supset \mathrm{C}_\pi^\infty$ with respect to the $\|\cdot\|_{\mathrm{L}^2}$ norm, we see that the map $\tilde{\mathcal{G}}$ can be uniquely extended[1] to a map

$$\mathcal{G} : \ell_2(\mathbb{Z}^d, \mathbb{C}) \to \mathrm{L}_2(\mathbb{T}^d, \mathbb{C}),$$

for which it holds that $\|\mathcal{G}(\hat{u})\|_{\mathrm{L}^2} = \|\hat{u}\|_{\ell_2}$. It is an exercise to show that this map is injective and surjective, so that $\mathcal{G}$ is an (isometric, i. e., norm preserving) isomorphism.[2] In the process, verify that the inverse $\mathcal{F} \in \mathcal{L}(\mathrm{L}^2(\mathbb{T}^d, \mathbb{C}), \ell_2(\mathbb{Z}^d, \mathbb{C}))$ of $\mathcal{G}$ is given by

$$\mathcal{F}(u) = \left(\frac{1}{(2\pi)^{d/2}} \int_{\mathbb{T}^d} u(x) e^{-ikx}\, dx \right)_{k\in\mathbb{Z}^d} = \big((u|\varphi_k)\big)_{k\in\mathbb{Z}^d}.$$

This map $\mathcal{F}$ is usually called *Fourier transform* and it holds that

$$u = \sum_{k\in\mathbb{Z}^d} \hat{u}_k \varphi_k \quad \text{provided } \hat{u}_k = (u|\varphi_k) \text{ for } k \in \mathbb{Z}^d.$$

This is compatible with the standard notation $\hat{u} = \mathcal{F}(u)$ that we shall adopt from now on. If we take $u \in \mathrm{L}^2(\mathbb{T}^d)$, we can imagine that

$$\Delta u = \Delta \sum_{k\in\mathbb{Z}^d} \hat{u}_k \varphi_k = \sum_{k\in\mathbb{Z}^d} \hat{u}_k \Delta\varphi_k = -\sum_{k\in\mathbb{Z}^d} |k|_2^2 \hat{u}_k \varphi_k.$$

This shows that, unless $(|k|_2^2 \hat{u}_k)_{k\in\mathbb{Z}^d} \in \ell_2(\mathbb{Z}^d, \mathbb{C})$, we may have convergence issues for the series, even though $\hat{u} \in \ell_2(\mathbb{Z}^d, \mathbb{C})$. In order to be able to better understand the mapping properties of Δ, we introduce additional spaces

$$\mathrm{H}_\pi^s = \left\{ u = \sum_{k\in\mathbb{Z}^d} \hat{u}_k \varphi_k \;\middle|\; \sum_{k\in\mathbb{Z}^d} \left[1 + |k|_2^2\right]^s |\hat{u}_k|^2 < \infty \right\}$$

for $s \in \mathbb{R}$ with norm given by

$$\|u\|_{s,2} = \sqrt{\sum_{k\in\mathbb{Z}^d} \left[1 + |k|_2^2\right]^s |\hat{u}_k|^2}.$$

---

1 This can be done using the fact of Section 4.3 and the fact that linear continuous maps are necessarily uniformly continuous.

2 For the purposes of this book, an *isomorphism* is a bijective, linear, and continuous map between normed vector spaces with a continuous inverse.

A somewhat more rigorous way to introduce these spaces is via completion of $\mathcal{F}^{-1}(c_{00}(\mathbb{Z}^d, \mathbb{C}))$ with respect to the norm $\|\cdot\|_{s,2}$. In any event,

$$\mathcal{F}(\mathrm{H}^s_\pi) = \ell^s_2(\mathbb{Z}^d, \mathbb{C}) = \Big\{\xi \in \mathbb{C}^{\mathbb{Z}^d} \;\Big|\; \sum_{k\in\mathbb{Z}^d} [1 + |k|_2^2]^s |\xi_k|^2 < \infty\Big\}.$$

Notice that $\mathrm{H}^0_\pi = \mathrm{L}^2(\mathbb{T}^d)$ and that its elements do not need to be periodic (nor continuous). Verify (Q2) the latter statement by showing that

$$h \in \mathrm{L}^2(\mathbb{T}) \quad \text{for } h(x) = \begin{cases} -1, & x \in [0, \pi), \\ 1, & x \in [\pi, 2\pi). \end{cases}$$

Show that $\mathrm{H}^s_\pi \subset \mathrm{H}^r_\pi$ for $r > s$ and that $\|u\|_{r,2} \leqslant C\|u\|_{s,2}$ for some constant $C > 0$ (which depends on r, s). The inequality shows that the inclusion map is continuous. When this is the case, one speaks of an *embedding* of spaces. The parameter $s \in \mathbb{R}$ measures the regularity of the elements of H^s_π. Indeed, we have the following.

Fact. *For any $s > d/2$, it holds that $u \in \mathrm{C}(\mathbb{T}^d)$ for any $u \in \mathrm{H}^s_\pi$. In particular, any such function is periodic in the classical sense.*

Fix $x_0 \in \mathbb{R}^d$ and define the linear functional δ_{x_0} by

$$\langle \delta_{x_0}, \psi\rangle = \psi(x_0) \quad \text{for } \psi \in \mathrm{C}^\infty_\pi,$$

and notice that

$$\langle \delta_{x_0}, \varphi_k\rangle = \frac{1}{(2\pi)^{d/2}} e^{-ikx_0} = (\widehat{\delta_{x_0}})_k = \widehat{\delta}^k_{x_0}, \quad k \in \mathbb{Z}^d.$$

It follows that

$$\psi(x_0) = \Big\langle \delta_{x_0}, \sum_{k\in\mathbb{Z}^d} \widehat{\psi}_k \varphi_k \Big\rangle = \sum_{k\in\mathbb{Z}^d} \widehat{\psi}_k \widehat{\delta}^k_{x_0}, \quad \widehat{\psi} \in c_{00}(\mathbb{Z}^d, \mathbb{C}),$$

and hence

$$\delta_{x_0}|_{\mathrm{L}_{00}} = \sum_{k\in\mathbb{Z}^d} \widehat{\delta}^k_{x_0} \varphi_k,$$

where

$$\mathrm{L}_{00} = \Big\{\sum_{k\in\mathbb{Z}^d} \xi_k \varphi_k \;\Big|\; \xi \in c_{00}(\mathbb{Z}^k, \mathbb{C})\Big\}.$$

Moreover, it holds that

$$\sum_{k\in\mathbb{Z}^d} [1+|k|_2^2]^s \hat{\delta}_{x_0}^k = \frac{1}{(2\pi)^{d/2}} \sum_{k\in\mathbb{Z}^d} [1+|k|_2^2]^s < \infty \quad \text{if } s < -d/2.$$

This shows that $\delta_{x_0} \in \mathrm{H}_\pi^{-s}$ for $s > d/2$. The Cauchy–Schwarz inequality implies that

$$\langle \delta_{x_0}, \psi \rangle = \sum_{k\in\mathbb{Z}^d} \hat{\delta}_{x_0}^k \hat{\psi}_k \frac{(1+|k|_2^2)^s}{(1+|k|_2^2)^s} \leqslant \|\delta_{x_0}\|_{-s,2} \|\psi\|_{s,2},$$

for $\psi \in \mathrm{L}_{00}$, and hence for $\psi \in \mathrm{H}_\pi^s$ by density.[3] With this in hand, we see that

$$|\psi(x) - \psi(y)| = |\langle \delta_x, \psi\rangle - \langle \delta_y, \psi\rangle| = |\langle \delta_x - \delta_y, \psi\rangle| \leqslant \|\delta_x - \delta_y\|_{-s,2} \|\psi\|_{s,2},$$

where

$$\begin{aligned}\|\delta_x - \delta_y\|_{-s,2}^2 &= \frac{1}{(2\pi)^{d/2}} \sum_{k\in\mathbb{Z}^d} |e^{-ikx} - e^{-iky}|^2 [1+|k|_2^2]^{-s} \\ &\leqslant \frac{4^{1-\varepsilon}}{(2\pi)^d} \sum_{k\in\mathbb{Z}^d} \frac{|k|_2^{2\varepsilon}}{(1+|k|_2^2)^s} |x-y|^{2\varepsilon} = C\,|x-y|^{2\varepsilon}, \quad x, y \in \mathbb{R}^d,\end{aligned}$$

provided $s > d/2 + \varepsilon$ for some $\varepsilon > 0$, since

$$|e^{-ikx} - e^{-iky}| \leqslant 2 \quad \text{and} \quad |e^{-ikx} - e^{-iky}| \leqslant |k|_2 |x-y|_2, \quad x, y \in \mathbb{R}^d.$$

This gives continuity and we leave it to the reader to show that periodicity follows. √

Observing (and you are encouraged to provide a formal proof) that

$$\partial_j \in \mathcal{L}(\mathrm{H}_\pi^s, \mathrm{H}_\pi^{s-1}) \quad \text{for } j = 1, \dots, d,$$

i. e., that differentiation in any direction is a linear and continuous (bounded) operation between the given spaces for any $s \in \mathbb{R}$, we use the above fact to show that

$$u \in \mathrm{H}_\pi^s \Longrightarrow u \in \mathrm{C}_\pi^m,$$

for $m \in \mathbb{N}$ if $s > m + d/2$, and, in this case, that

$$\partial^\alpha u(0) = \partial^\alpha u(2\pi e_j) \quad \text{for } j = 1, \dots, d \text{ and } |\alpha| \leqslant m,\ \alpha \in \mathbb{N}^d.$$

It follows that we also have the inclusion

$$\bigcap_{s\in\mathbb{R}} \mathrm{H}_\pi^s \subset \mathrm{C}_\pi^\infty .$$

---

3 Density means that any function in H_π^s can be approximated arbitrarily well by elements of L_{00} in the $\|\cdot\|_{s,2}$ norm.

Notice that, in a way fully analogous to the case of linear operators between finite-dimensional spaces (i. e., essentially, of matrices), we can write

$$-\Delta = \sum_{k\in\mathbb{Z}^d} |k|_2^2 \varphi_k \overline{\varphi_k}^{\top},$$

if we set $\overline{\varphi_k}^{\top} = (\cdot|\varphi_k) = \int_{\mathbb{T}^d} \cdot(x)\overline{\varphi_k}(x)\,dx$. The fact that the sum is infinite and that the set of eigenvalues is not bounded introduces, however, a new phenomenon: the linear operator $-\Delta$ is not bounded. To be more precise, we can try to define $-\Delta$ on H^s_π, in which case you should verify (Q3) that

$$-\Delta u \in \mathrm{H}^{s-2}_\pi \quad \text{for } u \in \mathrm{H}^s_\pi,$$

which entails (check)[4] that it is impossible to find a constant $C > 0$ such that

$$\|\Delta u\|_{s,2} \leqslant C\|u\|_{s,2},$$

which would amount to boundedness. This means that Δ, considered as an operator on H^s_π can only be defined for functions in H^{s+2}_π, which is its *domain*. We write this as

$$\Delta : \mathrm{dom}(\Delta) = \mathrm{H}^{s+2}_\pi \subset \mathrm{H}^s_\pi \to \mathrm{H}^s_\pi,$$

where the inclusion indicates that we consider H^{s+2}_π with the topology induced by H^s_π. We can alternatively consider the native topology on H^{s+2}_π, in which case we would actually have a bounded operator (show it), i. e., $\Delta \in \mathcal{L}(\mathrm{H}^{s+2}_\pi, \mathrm{H}^s_\pi)$. This would, however, not be convenient since we would like to consider the periodic heat equation $u_t = \Delta u$ as an infinite-dimensional ODE. This requires that we interpret Δ as a "vector field" $V : E \to E$ on some space E. At a purely formal level, it is straightforward to see that

$$u(t, u_0) = \sum_{k\in\mathbb{Z}^d} e^{-t|k|_2^2} \hat{u}_{0k}\varphi_k, \quad t > 0$$

is a solution of the periodic heat equation with initial datum

$$u_0 = \sum_{k\in\mathbb{Z}^d} \hat{u}_{0k}\varphi_k.$$

Fact. *Given $u_0 \in \mathrm{H}^s_\pi$ for $s \in \mathbb{R}$, it holds that*

$$u(\cdot, u_0) = \sum_{k\in\mathbb{Z}^d} e^{-\cdot|k|_2^2} \hat{u}_{0k}\varphi_k \in \mathrm{C}([0,\infty), \mathrm{H}^s_\pi)$$

4 Construct a sequence $(u_n)_{n\in\mathbb{N}}$ of functions in H^s_π such that $\|\Delta u_n\|_{s,2} = n^2\|u_n\|_{s,2}$.

solves the periodic heat equation in H^s_π *with initial datum* u_0*. Furthermore,*

$$u \in \mathrm{C}^m([0,\infty), \mathrm{H}^{s-2m}_\pi),$$

for $m \in \mathbb{N}$*.*

If $u_0 \in \mathrm{H}^s_\pi$, then

$$\sum_{k\in\mathbb{Z}^d} [1+|k|_2^2]^s |\widehat{u}_{0k}|^2 < \infty.$$

For $t \geqslant s \geqslant 0$, we have that

$$\begin{aligned}\|u(t,u_0) - u(s,u_0)\|^2_{\mathrm{H}^s_\pi} &= \Big\| \sum_{k\in\mathbb{Z}^d} (e^{-t|k|_2^2} - e^{-s|k|_2^2})\widehat{u}_{0k}\varphi_k \Big\|_{\mathrm{H}^s_\pi} \\ &= \sum_{k\in\mathbb{Z}^d} [1+|k|_2^2]^s (e^{-t|k|_2^2} - e^{-s|k|_2^2})^2 |\widehat{u}_{0k}|^2.\end{aligned}$$

For $\varepsilon > 0$, $N_\varepsilon \in \mathbb{N}$ can be found such that

$$\begin{aligned}\sum_{|k|_2\geqslant N} [1+|k|_2^2]^s (e^{-t|k|_2^2} - e^{-s|k|_2^2})^2 |\widehat{u}_{0k}|^2 &\leqslant 4 \sum_{|k|_2\geqslant N_\varepsilon} [1+|k|_2^2]^s |\widehat{u}_{0k}|^2 \\ &\leqslant \varepsilon^2/2 \quad \text{for } N \geqslant N_\varepsilon,\end{aligned}$$

since $u_0 \in \mathrm{H}^s_\pi$. Then there is $\delta > 0$ such that

$$\sum_{|k|_2<N_\varepsilon} [1+|k|_2^2]^s (e^{-t|k|_2^2} - e^{-s|k|_2^2})^2 |\widehat{u}_{0k}|^2 \leqslant \varepsilon^2/2 \quad \text{for } |t-s| \leqslant \delta,$$

by continuity, since the sum is finite. Combining the inequalities, we see that: For any given $\varepsilon > 0$, there is $\delta > 0$ such that

$$\|u(t,u_0) - u(s,u_0)\|_{\mathrm{H}^s_\pi} \leqslant \varepsilon \quad \text{for } |t-s| \leqslant \delta,$$

and the claimed continuity follows.

Observe that

$$u^{(m)}(t,u_0) = \sum_{k\in\mathbb{Z}^d} (-|k|_2^2)^m e^{-t|k|_2^2} \widehat{u}_{0k}\varphi_k.$$

As $(|k|_2^{2m}\widehat{u}_{0k})_{k\in\mathbb{Z}^d}$ satisfies

$$\sum_{k\in\mathbb{Z}^d} |k|_2^{4m} [1+|k|_2^2]^{s-2m} |\widehat{u}_{0k}|^2 \leqslant \sum_{k\in\mathbb{Z}^d} [1+|k|_2^2]^{2m} [1+|k|_2^2]^{s-2m} |\widehat{u}_{0k}|^2 < \infty,$$

since $u_0 \in \mathrm{H}^s_\pi$, we see that

$$u^{(m)}(t, u_0) = \Delta^m u(t, u_0) \in \mathrm{H}^{s-2m}_\pi, \quad t \geqslant 0,$$

and continuity follows as in the first part of this proof. √

Defining

$$T(t)u_0 = \sum_{k \in \mathbb{Z}^d} e^{-t|k|_2^2} \hat{u}_{0k} \varphi_k = \text{“}e^{t\Delta}u_0\text{”}, \quad u_0 \in \mathrm{H}^s_\pi,$$

we obtain a so-called *strongly continuous semigroup* on H^s_π for any $s \in \mathbb{R}$, which amounts to the validity of:

(i) $T(t+s) = T(t)T(s)$ for $t, s \geqslant 0$.
(ii) $T(0) = \mathrm{id}_{\mathrm{H}^s_\pi}$.
(iii) $T(\cdot)u_0 \in \mathrm{C}([0,\infty), \mathrm{H}^s_\pi)$ $\forall u_0 \in \mathrm{H}^s_\pi$.

The reader is urged to verify that

$$u(t, u_0) = T(t)u_0 \in \bigcap_{s \in \mathbb{R}} \mathrm{H}^s_\pi \quad \text{for } t > 0,$$

so that $T(t)u_0 \in \mathrm{C}^\infty_\pi$ for $t > 0$, and consequently, that

$$T(\cdot)u_0 \in \mathrm{C}^\infty((0,\infty), \mathrm{C}^\infty_\pi).$$

Also verify that

$$\|\Delta T(t)u_0\|_{\mathrm{H}^s_\pi} \leqslant \frac{C}{t} \|u_0\|_{\mathrm{H}^s_\pi}, \quad t > 0,$$

which shows that regularity does not hold up to $t = 0$, in general. Finally, convince yourself that, in stark contrast to ODEs in finite dimensions (or in infinite dimensions but for a bounded operator), the equation cannot be solved for negative times for generic initial data.

9.2 The heat equation on the whole space

Next, we consider the heat equation on the whole space $\mathbb{R}^d$,

$$u_t = \Delta u, \quad u(0, \cdot) = u_0.$$

In analogy with the periodic case, we consider the *Fourier transform* defined by

$$\hat{u}(\xi) = \mathcal{F}(u)(\xi) = \frac{1}{(2\pi)^{d/2}} \int_{\mathbb{R}^d} e^{-ix\cdot\xi} u(x)\, dx, \quad \xi \in \mathbb{R}^n,$$

for

$$u \in \mathcal{D}(\mathbb{R}^d) = \mathcal{D} = \{u \in C^\infty(\mathbb{R}^d, \mathbb{C}) \mid \operatorname{supp}(u) = \overline{[u \neq 0]} \text{ is compact}\}.$$

Here, the *support* of a function is defined as the closure of the set on which the function does not vanish. Observing that

$$|\mathcal{F}(u)(\xi)| \leqslant C \int_{\mathbb{R}^d} \underbrace{|e^{-ix\cdot\xi}|}_{=1} |u(x)|\, dx = C\|u\|_1, \quad \xi \in \mathbb{R}^d \text{ for } u \in \mathcal{D},$$

we see that the definition can be extended to all functions in the completion of $\mathcal{D}$ with respect to the $\|\cdot\|_1$-norm

$$L^1(\mathbb{R}^d) = \overline{\mathcal{D}}^{\|\cdot\|_1}.$$

The motivation for the use of the Fourier transform, as in the periodic case, is its ability to diagonalize differentiation, i. e., (check)

$$\mathcal{F}(\partial_j u) = (i\xi_j)\mathcal{F}(u) \quad \text{for } u \in \mathcal{D} \text{ and } j = 1, \dots, d,$$

which can also be written as $\partial_j = \mathcal{F}^{-1} \circ M_{i\xi_j} \circ \mathcal{F}$ once we prove that $\mathcal{F}$ is invertible in well-chosen function spaces. We use the notation M_f for the linear operator of pointwise multiplication given by $[u \mapsto fu]$. Taking a Fourier transform of the heat equation yields

$$\hat{u}_t = -|\xi|_2^2 \hat{u}, \quad \hat{u}(t, \cdot) = \hat{u}_0,$$

which can be solved to give

$$\mathcal{F}(u)(t, \xi) = e^{-t|\xi|_2^2} \mathcal{F}(u_0)(\xi), \quad t > 0,\ \xi \in \mathbb{R}^d.$$

Once invertibility of $\mathcal{F}$ is well understood, this can be used to obtain a solution of the heat equation. We therefore study the Fourier transform in more detail. Exploiting the fact that

$$e^{-ix\cdot\xi_n} u(x) \longrightarrow e^{-ix\xi} u(x) \quad \text{uniformly in } x \in \mathbb{R}^d \text{ as } n \to \infty,$$

if $\xi_n \to \xi$ as $n \to \infty$ for any fixed $u \in \mathcal{D}$ (why? (Q4)) we see that $\hat{u}$ is a continuous function, which is also bounded since

$$\|\hat{u}\|_\infty = \sup_{\xi \in \mathbb{R}^d} |\hat{u}(\xi)| \leqslant C\|u\|_1$$

as shown above. In search of spaces on which the Fourier transform may be invertible and motivated by the fact that it turns derivatives into multiplications by monomials and multiplication by monomials into derivatives as follows from

$$\begin{aligned}\mathcal{F}(M_{x_j}u) = \mathcal{F}(x_j u) &= \frac{1}{(2\pi)^{d/2}} \int_{\mathbb{R}^d} x_j e^{-ix\cdot\xi} u(x) = \frac{i}{(2\pi)^{d/2}} \int_{\mathbb{R}^d} \partial_j e^{-ix\cdot\xi} u(x) \\ &= i\partial_j \frac{1}{(2\pi)^{d/2}} \int_{\mathbb{R}^d} e^{-ix\cdot\xi} u(x) = i\partial_j \mathcal{F}(u), \quad j = 1, \ldots, d,\ u \in \mathcal{D},\end{aligned}$$

we introduce the space

$$\mathcal{S} = \Big\{u \in \mathrm{C}^\infty(\mathbb{R}^d, \mathbb{C}) \mid \forall m, k \in \mathbb{N}\ q_{m,k}(\varphi) = \sup_{x\in\mathbb{R}^d} \sup_{|\alpha|\leqslant m} \big[1 + |x|_2^2\big]^{k/2} \big|\partial^\alpha u(x)\big| < \infty\Big\},$$

which is invariant with respect to taking derivatives and to multiplying by polynomials. Given $u \in \mathcal{S}$ and $m, k \in \mathbb{N}$, a constant $C = C(m,k)$ can be found such that

$$\big|\partial^\alpha u(x)\big| \leqslant C \frac{C}{[1 + |x|_2^2]^{k/2}}, \quad x \in \mathbb{R}^d,$$

for any $\alpha \in \mathbb{N}^d$ such that $|\alpha| \leqslant m$. Convince yourself that any partial derivative is integrable, i.e. that $\partial^\alpha u \in \mathrm{L}^1(\mathbb{R}^d)$ for any $\alpha \in \mathbb{N}^d$ if $u \in \mathcal{S}$, and show that

$$\mathcal{F}(\partial^\alpha u) = (i\xi)^\alpha \mathcal{F}(u), \quad \mathcal{F}(x^\alpha u) = i^{|\alpha|} \partial^\alpha \mathcal{F}(u), \quad u \in \mathcal{D}.$$

It is a tedious but instructive exercise to verify that

$$\hat{u} \in \mathcal{S} \Longleftrightarrow u \in \mathcal{S}.$$

The function $\psi_e(x) = e^{-\frac{|x|_2^2}{2}}$, $x \in \mathbb{R}^d$, is an element of $\mathcal{S}$, which does not belong to $\mathcal{D} \subset \mathcal{S}$. Notice

$$\int_{\mathbb{R}^d} e^{-ix\cdot\xi} e^{-\frac{|x|_2^2}{2}}\, dx = e^{-\frac{|\xi|_2^2}{2}} \int_{\mathbb{R}^d} e^{-(x+i\xi)^2/2}\, d(x + i\xi) = (2\pi)^{d/2} e^{-\frac{|\xi|_2^2}{2}},$$

so that $\mathcal{F}(\psi_e) = \psi_e$. If σ_s denotes dilation by s, that is, if we have that $\sigma_s(u) = u(s\,\cdot)$ for a function $u : \mathbb{R}^d \to \mathbb{C}$, then

$$\begin{aligned}\mathcal{F}(\sigma_s(u)) &= \frac{1}{(2\pi)^{d/2}} \int_{\mathbb{R}^d} e^{-ix\xi} u(sx)\, dx = \frac{1}{(2\pi)^{d/2}} \frac{1}{s^d} \int_{\mathbb{R}^d} e^{-ix\xi/s} u(x)\, dx \\ &= \frac{1}{s^d} \sigma_{1/s}(\mathcal{F}(u)).\end{aligned}$$

Given functions $u, v \in \mathcal{S}$, we define their *convolution* $u * v$ by

$$(u * v)(x) = \int_{\mathbb{R}^d} u(x-y)v(y)\,dy.$$

Show that $u * v \in \mathcal{S}$. Using the function ψ_e introduced above, define $\varphi = \frac{\psi_e}{\|\psi_e\|_1}$ and set $\varphi_\varepsilon = \frac{1}{\varepsilon^d}\sigma_{1/\varepsilon}(\varphi)$, so that $\int_{\mathbb{R}^d} \varphi_\varepsilon(x)\,dx = 1$ for $\varepsilon > 0$.

Fact (Approximation of the identity). *It holds that*

$$\varphi_\varepsilon * u \longrightarrow u \quad as\ \varepsilon \to 0,$$

where the convergence is in the $\|\cdot\|_p$*-norm if* $u \in \mathrm{L}^p(\mathbb{R}^d)$ *and uniformly on compact subsets if u is bounded and continuous. More explicitly, it holds that*

$$\int_{\mathbb{R}^d} |\varphi_\varepsilon * u(x) - u(x)|^p\,dx \to 0,$$

in the first case, and that $\sup_{x\in K} |\varphi_\varepsilon * u(x) - u(x)| \to 0$ *for any compact* $K \subset \mathbb{R}^d$*, in the second case, as* $\varepsilon \to 0$.

We start with the second claim. First, observe that

$$\int_{\mathbb{R}^d} \varphi_\varepsilon(x-y)u(y)\,dy = \int_{\mathbb{R}^d} \varphi(z)u(x-\varepsilon z)\,dz,$$

as follows by changing variables in the integral ($z = \frac{x-y}{\varepsilon}$). Then

$$|\varphi_\varepsilon * u(x) - u(x)| = \left|\int_{\mathbb{R}^d} \varphi(z)\big[u(x-\varepsilon z) - u(x)\big]\,dz\right|.$$

Due to the exponential decay of ψ_e, a constant N_δ can be found such that

$$\int_{|z|\geqslant N_\delta} \varphi(z)\,dz \leqslant \frac{\delta}{4\|u\|_\infty},$$

for any given $\delta > 0$. Moreover, if x is taken from a compact subsets of $\mathbb{R}^d$, there is $\varepsilon_\delta > 0$ such that

$$\big|u(x-\varepsilon z) - u(x)\big| \leqslant \frac{\delta}{2} \quad \text{for } |z| \leqslant N_\delta, \text{ if } \varepsilon \leqslant \varepsilon_\delta.$$

We conclude that

$$|\varphi_\varepsilon * u(x) - u(x)| \leqslant \int_{|z|\geqslant N_\delta} \varphi(z)2\|u\|_\infty\,dz + \int_{|z|\leqslant N_\delta} \frac{\delta}{2}\varphi(z)\,dz \leqslant \delta,$$

provided $\varepsilon \leqslant \varepsilon_\delta$ since $\int_{\mathbb{R}^d} \varphi(z)\,dz = 1$ and $\varphi \geqslant 0$.

As for the first claim, we prove it for $p = 1$ and leave the case $p > 1$ as an exercise. The space $\mathcal{D}$ is dense in $L^1(\mathbb{R}^d)$ by definition. Thus, given $u \in L^1(\mathbb{R}^d)$ and $\delta > 0$, we can find $v \in \mathcal{D}$ such that $\|u - v\|_1 \leqslant \frac{\delta}{3}$. It follows that:

$$\begin{aligned}\|u * \varphi_\varepsilon - u\|_1 &\leqslant \int_{\mathbb{R}^d}\left|\int_{\mathbb{R}^d} \varphi(z)[u(x-\varepsilon z) - u(x)]\,dz\right| dx \\ &\leqslant \int_{\mathbb{R}^d} \varphi(z) \int_{\mathbb{R}^d} \{|u(x-\varepsilon z) - v(x-\varepsilon z)| \\ &\quad + |v(x-\varepsilon z) - v(x)| + |v(x) - u(x)|\}\,dxdz,\end{aligned}$$

and, since

$$\int_{\mathbb{R}^d} \varphi(z) \int_{\mathbb{R}^d} |u(x-\varepsilon z) - v(x-\varepsilon z)|\,dxdz = \int_{\mathbb{R}^d} \varphi(z) \int_{\mathbb{R}^d} |u(x) - v(x)|\,dxdz = \|u - v\|_1 \leqslant \frac{\delta}{3},$$

that

$$\|u * \varphi_\varepsilon - u\|_1 \leqslant \frac{2\delta}{3} + \int_{\mathbb{R}^d} \varphi(z) \int_{\mathbb{R}^d} |v(x-\varepsilon z) - v(x)|\,dxdz.$$

Arguing as in the proof of the second claim and using the fact that v is compactly supported (and hence also uniformly continuous), it is shown that there is $\varepsilon_\delta > 0$ such that

$$\int_{\mathbb{R}^d} \varphi(z) \int_{\mathbb{R}^d} |v(x-\varepsilon z) - v(x)|\,dxdz \leqslant \frac{\delta}{3} \int_{\mathbb{R}^d} \varphi(z)\,dz = \frac{\delta}{3},$$

for $\varepsilon \leqslant \varepsilon_\delta$, which concludes the argument. √

Fact (Convolution theorem). *It holds that*

$$\mathcal{F}(u * v) = (2\pi)^{d/2}\mathcal{F}(u)\mathcal{F}(v) \quad \text{for } u, v \in \mathcal{S}.$$

The verification of this identity is left as an exercise.

Fact (Inversion formula). *For any $u \in \mathcal{S}$, it holds that*

$$u(x) = \frac{1}{(2\pi)^{d/2}} \int_{\mathbb{R}^d} e^{ix\cdot\xi}\hat{u}(\xi)\,d\xi, \quad x \in \mathbb{R}^d.$$

We observe that

$$I(t) = \frac{1}{(2\pi)^{d/2}} \int_{\mathbb{R}^d} e^{ix\cdot\xi} e^{-t\frac{|\xi|_2^2}{2}} \hat{u}(\xi)\,d\xi$$

$$= \frac{1}{(2\pi)^d} \int_{\mathbb{R}^d} e^{ix\cdot\xi} \int_{\mathbb{R}^d} e^{-iy\cdot\xi} u(y)\, dy\, e^{-t\frac{|\xi|_2^2}{2}}\, dx$$

$$= \frac{1}{(2\pi)^d} \int_{\mathbb{R}^d}\int_{\mathbb{R}^d} e^{i(x-y)\cdot\xi} e^{-t\frac{|\xi|_2^2}{2}}\, d\xi\, u(y)\, dy$$

$$= \frac{1}{(2\pi)^{d/2}} \int_{\mathbb{R}^d} \frac{1}{t^{d/2}} e^{-\frac{|x-y|_2^2}{2t}} u(y)\, dy = r(t).$$

Now, notice that

$$l(t) \longrightarrow \frac{1}{(2\pi)^{d/2}} \int_{\mathbb{R}^d} e^{ix\cdot\xi} \hat{u}(\xi)\, d\xi,$$

as $t \searrow 0$, while $r(t) \to u(x)$ in the same limit thanks to

$$\frac{1}{(2\pi)^{d/2}} \frac{1}{t^{d/2}} e^{-\frac{|x|_2^2}{2t}} = \left(\frac{1}{\sqrt{t}}\right)^d \sigma_{\frac{1}{\sqrt{t}}}(\psi_e),$$

and the approximation of the identity property. This yields the inversion formula. √

Notice that we introduced the term $e^{-t\frac{|\xi|_2^2}{2}}$ to ensure convergence of the integrals in all steps of our calculation before letting the parameter t vanish. Performing a purely formal calculation with $t = 0$, would correspond to having that

$$\frac{1}{(2\pi)^{d/2}} \int_{\mathbb{R}^d} e^{i(x-y)\cdot\xi}\, d\xi = \delta(x-y),$$

with δ defined (again formally) by $\int_{\mathbb{R}^d} \delta(x)u(x)\, dx = u(0)$,[5] which follows from the fact that:

$$\mathcal{F}(\delta) = \frac{1}{(2\pi)^{d/2}} \int_{\mathbb{R}^d} e^{-ix\cdot\xi} \delta(x)\, dx = \frac{1}{(2\pi)^{d/2}},$$

by formally applying the inversion formula. Keeping this in mind as a guiding principle, give a proof of the following.

Fact (Plancherel). *It holds that*

$$\int_{\mathbb{R}^d} u(x)\overline{v}(x)\, dx = \int_{\mathbb{R}^d} \mathcal{F}(u)(\xi)\overline{\mathcal{F}(v)}(\xi)\, d\xi$$

for $u, v \in \mathcal{S}$. In particular, we have that

5 This can be justified by properly defining δ as a measure or as a generalized function.

$$\|u\|_2 = \left(\int_{\mathbb{R}^d} |u(x)|^2\,dx\right)^{\frac{1}{2}} = \left(\int_{\mathbb{R}^d} |\hat{u}(x)|^2\,dx\right)^{\frac{1}{2}} = \|\mathcal{F}(u)\|_2$$

for $u \in \mathcal{S}$, i. e., the Fourier transform is an isometry on $\mathcal{S}$ with respect to the $\|\cdot\|_2$-norm.

Given the above fact and the density of $\mathcal{S} \supset \mathcal{D}$ in $L^2(\mathbb{R}^d)$,[6] we can extend the Fourier transform to $L^2(\mathbb{R}^d)$. Indeed, if $u \in L^2(\mathbb{R}^d)$, take a sequence $(\varphi_n)_{n\in\mathbb{N}}$ in $\mathcal{S}$ satisfying $\|\varphi_n - u\|_2 \to 0$ as $n \to \infty$, so that $(\varphi_n)_{n\in\mathbb{N}}$ is a Cauchy sequence. Then

$$\|\widehat{\varphi}_n - \widehat{\varphi}_m\|_2 = \|\varphi_n - \varphi_m\|_2 \to 0 \quad \text{as } m, n \to \infty$$

shows that $(\widehat{\varphi}_n)_{n\in\mathbb{N}}$ is also Cauchy, and as such, has a limit $v \in L^2(\mathbb{R}^d)$ by completeness. We define $v = \hat{u} = \mathcal{F}(u)$. Show that this is well-defined, i. e., that v is uniquely determined by u. The inverse $\mathcal{F}^{-1}$ can similarly be extended to yield an inverse for the extension of $\mathcal{F}$ (check this). We shall continue to use the notation $\mathcal{F}$ even for the extension, which is an isometric isomorphism.

We observe that, while one often writes

$$\mathcal{F}(u)(\xi) = \frac{1}{(2\pi)^{d/2}} \int_{\mathbb{R}^d} e^{-ix\cdot\xi} u(x)\,dx$$

for $u \in L^2(\mathbb{R}^d)$, the integral does not actually always exist for such u. You are asked to verify that, however, we have that

$$\left\| \mathcal{F}(u) - \frac{1}{(2\pi)^{d/2}} \int_{|x|_2 \leqslant R} e^{-ix\cdot(\cdot)} u(x)\,dx \right\|_2 \to 0 \quad \text{as } R \to \infty.$$

In a way similar to what we did for the periodic heat equation, we introduce the spaces

$$H^s(\mathbb{R}^d) = \{u \in L_2(\mathbb{R}^d) \mid [1 + |\xi|_2^2]^{s/2}\hat{u} \in L^2(\mathbb{R}^n)\}$$

for $s \geqslant 0$. These are Hilbert spaces with respect to the inner product given by

$$(u|v)_{H^s} = \int_{\mathbb{R}^d} \hat{u}(x)\bar{\hat{v}}(\xi)[1 + |\xi|_2^2]^s\,d\xi.$$

Notice that $H^0(\mathbb{R}^n) = L^2(\mathbb{R}^d)$. Spaces can be obtained also for $s < 0$ by defining $H^s(\mathbb{R}^d)$ as the completion of $\mathcal{S}$ (or of $L^2(\mathbb{R}^d)$) in the norm given by

$$\|u\|_{H^s} = \left(\int_{\mathbb{R}^d} |\hat{u}(\xi)|^2 [1 + |\xi|_2^2]^s\,d\xi\right)^{\frac{1}{2}}.$$

6 We define $L^p(\mathbb{R}^d)$ for $p \in (1, \infty)$ as the completion of $\mathcal{D}$ in the $\|\cdot\|_p$ norm.

These are also Hilbert spaces with the corresponding scalar product and it holds that $\mathrm{H}^s(\mathbb{R}^d) \subset \mathrm{H}^r(\mathbb{R}^d)$ if $s \geqslant r$. Notice that $\psi_e \in \mathrm{H}^s(\mathbb{R}^d)$ for each $s \in \mathbb{R}$ and that, similar to the periodic case, $\delta \in \mathrm{H}^s(\mathbb{R}^d)$ provided we have that $s < -d/2$ (convince yourself of this even if you cannot give a completely rigorous proof). It follows, again in a way fully parallel to the periodic case, that we have the following.

Fact. *For $s > d/2$, it holds that*

$$\mathrm{H}^s(\mathbb{R}^d) \subset \mathrm{BUC}(\mathbb{R}^d),$$

where

$$\mathrm{BUC}(\mathbb{R}^d) = \{u : \mathbb{R}^d \to \mathbb{C} \mid u \textit{ is bounded and uniformly continuous}\}.$$

This, together with the fact that

$$\partial_j \in \mathcal{L}(\mathrm{H}^s(\mathbb{R}^d), \mathrm{H}^{s-1}(\mathbb{R}^d)) \quad \textit{for } j = 1, \dots, d \textit{ and } s \in \mathbb{R},$$

implies that $\bigcap_{s>0} \mathrm{H}^s(\mathbb{R}^d) \subset \mathrm{C}^\infty(\mathbb{R}^d, \mathbb{C})$.

Returning to the heat equation, we will now verify that

$$H(t,x) = \frac{1}{(4\pi t)^{d/2}} e^{-\frac{|x|_2^2}{4t}}, \quad (t,x) \in (0,\infty) \times \mathbb{R}^d,$$

is a solution satisfying $u(0,\cdot) = \delta$ in the sense that

$$\int_{\mathbb{R}^d} H(t,\cdot - y)\varphi(y)\,dy \to \varphi \quad \text{in } \mathrm{L}^2(\mathbb{R}^d) \text{ as } t \searrow 0,$$

for $\varphi \in \mathrm{L}^2(\mathbb{R}^d)$ (and hence in the previously defined sense of an approximation of the identity). Before proving this claim, we make a parallel with the finite-dimensional case of a linear ODE fully explicit. The solution of $\dot{x} = Ax$, $x(0) = x_0$ for $x : [0,\infty) \to \mathbb{R}^n$, and $A \in \mathbb{R}^{n\times n}$ is given by $e^{tA}u_0$. If we are able to find/compute $e^{tA}v_k$ for $k = 1,\dots,n$ and a basis $v_1,\dots,v_n$, denote the coefficients of u_0 in this basis by u_0^k, i. e., $u_0 = \sum_{k=1}^n u_0^k v_k$, then we would have that

$$e^{tA}u_0 = e^{tA}\left(\sum_{k=1}^n u_0^k v_k\right) = \sum_{k=1}^n u_0^k e^{tA} v_k.$$

In other words, we can synthesize the solution to any initial datum by means of the solutions, which have the basis vectors as initial data. If you replace the basis vectors by $(\delta_y)_{y\in\mathbb{R}^d}$, where $\delta_y = \delta(\cdot - y)$ and can find/compute the solution of $u_t = \Delta u$, $u(0,\cdot) = \delta_y$ for $y \in \mathbb{R}^d$, i. e., $e^{t\Delta}\delta_y = H(t,\cdot - y)$, then the solution to a general initial datum

$$u_0(\cdot) = \int_{\mathbb{R}^d} u_0(y)\delta_y(\cdot)\,dy = \text{“}\sum_{y\in\mathbb{R}^d} u_0(y)\delta_y(\cdot)\text{”},$$

is given by

$$\begin{aligned} e^{t\Delta}u_0 &= e^{t\Delta}\int_{\mathbb{R}^d} \delta_y(\cdot)u_0(y)\,dy = \int_{\mathbb{R}^d} \underbrace{e^{t\Delta}\delta_y(\cdot)}_{=H(t,\cdot-y)}u_0(y)\,dy \\ &= \int_{\mathbb{R}^d} H(t,\cdot - y)u_0(y)\,dy. \end{aligned}$$

In contrast to the finite-dimensional case, more care needs to be taken in making sure that the integrals do make sense. This translates into conditions on the initial datum u_0, which cannot be just any function in $\mathbb{C}^{\mathbb{R}^d}$, whereas it can be any vector in $\mathbb{C}^n$ in the finite-dimensional case since the construction works for any $x_0 \in \mathbb{C}^n$.

In order to verify that H is indeed a solution of the heat equation, we let ourselves be guided by another pertinent analogy with the finite-dimensional case. For matrices and for the Laplacian Δ in the periodic case, we saw that eigenvalues/eigenvectors open the door to the effective representation of solutions. The functions

$$e_\xi(x) = \frac{1}{(2\pi)^{d/2}}e^{ix\cdot\xi}, \quad x \in \mathbb{R}^d,$$

parametrized by $\xi \in \mathbb{R}^d$ are formally eigenfunctions of Δ to the eigenvalue $-|\xi|_2^2$ and they formally satisfy

$$e^{t\Delta}e_\xi = e^{-t|\xi|_2^2}e_\xi.$$

Thus, if we can represent any initial datum u_0 as a “combination” of such functions,

$$u_0 = \frac{1}{(2\pi)^{d/2}}\int_{\mathbb{R}^d} \hat{u}_0(\xi)e^{ix\cdot\xi}\,d\xi = \int_{\mathbb{R}^d} \hat{u}_0(\xi)e_\xi\,d\xi = \text{“}\sum_{\xi\in\mathbb{R}^d} \hat{u}_0(\xi)e_\xi\text{”},$$

then we obtain a representation of the corresponding solution as

$$e^{t\Delta}u_0 = \frac{1}{(2\pi)^{d/2}}\int_{\mathbb{R}^d} e^{-t|\xi|_2^2}\hat{u}_0(\xi)e^{ix\cdot\xi}\,d\xi = \text{“}\sum_{\xi\in\mathbb{R}^d} e^{-t|\xi|_2^2}\hat{u}_0(\xi)e_\xi\text{”}.$$

This can now be justified by using the Fourier transform of the equation to obtain

$$\hat{u}_t = |\xi|_2^2\hat{u}, \quad \hat{u}(0,\cdot) = \hat{\delta} = \frac{1}{(2\pi)^{d/2}},$$

solving the (parameter dependent) ODE to get

$$\widehat{u}(t,\xi) = e^{-t|\xi|_2^2}\widehat{u}_0 = \frac{1}{(2\pi)^{d/2}}e^{-t|\xi|_2^2} = \frac{1}{(2\pi)^{d/2}}e^{-|\sqrt{2t}\xi|_2^2/2},$$

and using the inversion formula, the dilation property of the Fourier transform, and $\mathcal{F}\psi_e = \psi_e$ to finally see that

$$\begin{aligned} u(t,x) &= \frac{1}{(\sqrt{2t})^d}\frac{1}{(2\pi)^{d/2}}e^{-|\frac{x}{\sqrt{2t}}|_2^2/2} = \frac{1}{(4t\pi)^{d/2}}e^{-\frac{|x|_2^2}{4t}} \\ &= H(t,x), \quad (t,x) \in (0,\infty) \times \mathbb{R}^d. \end{aligned}$$

Using the spaces $\mathrm{H}^s(\mathbb{R}^d)$ and the properties of the Fourier transform, we can get precise regularity statements about solutions of the heat equation, the proofs of which are left as an exercise. If $u_0 \in \mathrm{H}^s(\mathbb{R}^d)$, then the solution satisfies

$$u \in \mathrm{C}^m([0,\infty), \mathrm{H}^{s-2m}(\mathbb{R}^d)) \quad \text{for } m = 0, 1, \ldots$$

In particular, if $u_0 \in \mathrm{H}^2(\mathbb{R}^d)$, then

$$u \in \mathrm{C}([0,\infty), \mathrm{H}^2(\mathbb{R}^d)) \cap \mathrm{C}^1([0,\infty), \mathrm{L}^2(\mathbb{R}^d)).$$

The strongly continuous semigroup on $\mathrm{L}^2(\mathbb{R}^d)$ (check) defined by

$$T(t)u_0 = e^{t\Delta}u_0 = (\mathcal{F}^{-1} \circ M_{e^{-t|\cdot|_2^2}} \circ \mathcal{F})(u_0) = H(t,\cdot) * u_0$$

satisfies

$$T(t)(\mathrm{L}^2(\mathbb{R}^d)) \subset \mathrm{H}^s(\mathbb{R}^d), \quad t > 0, \text{ for any } s \geqslant 0,$$

so that $T(t)(\mathrm{L}^2(\mathbb{R}^d)) \subset \mathrm{C}^\infty(\mathbb{R}^d)$ for $t > 0$, and consequently,

$$T(\cdot)u_0 \in \mathrm{C}^\infty((0,\infty), \mathrm{C}^\infty(\mathbb{R}^d)).$$

Moreover, it holds that

$$\|T(t)u_0\|_{\mathrm{H}^s} \leqslant \frac{C}{t^{s/2}}\|u_0\|_{\mathrm{H}^s}, \quad t > 0,$$

showing that the regularity does not hold up to $t = 0$ in general.

We conclude this section by a brief and incomplete discussion of the inhomogeneous heat equation

$$u_t = \Delta u + f(t), \quad u(0) = u_0,$$

which, for given $f : [0,T] \to \mathrm{H}^s(\mathbb{R}^d)$ and some $s \in \mathbb{R}$, we think of as an equation for $u : [0,T] \to \mathrm{H}^s(\mathbb{R}^d)$. Given the functional framework, we developed and properties of the Fourier transform. This equation can be thought of as an infinite-dimensional ODE with formal solution given by

$$u(t, u_0) = e^{t\Delta} u_0 + \int_0^t e^{(t-\tau)\Delta} f(\tau)\, d\tau, \quad t \in [0, T].$$

It is an excellent exercise to think about what properties f may need to possess in order to show that the formula above does actually make sense, first, and yields an actual solution of the inhomogeneous equation, second. It is advisable to start by considering $s = 0$ and by taking a Fourier transform in the spatial variable.

9.3 Concluding remarks

9.3.1 Infinite-dimensional optimization

Introducing the quadratic functional

$$E(u) = \frac{1}{2} \int_{\mathbb{R}^n} |\nabla u|^2\, dx, \quad u \in \mathrm{H}^1(\mathbb{R}^d),$$

noticing that it is differentiable with

$$DE(u)h = \int \nabla u \cdot \nabla h\, dx \quad \text{for } h \in \mathrm{H}^1(\mathbb{R}^d),$$

it is possible to interpret the heat/diffusion equation as a gradient flow. If $u \in \mathrm{H}^2(\mathbb{R}^d)$, integration by parts[7] yields

$$\begin{aligned} DE(u)h &= \sum_{j=1}^d \int \partial_j u\, \partial_j h\, dx = \int \Big(- \sum_{j=1}^d \partial_j^2 u \Big) h\, dx \\ &= \int (-\Delta u) h\, dx = (-\Delta u | h)_2, \quad h \in \mathrm{H}^1(\mathbb{R}^d). \end{aligned}$$

This means that $\nabla E(u) = -\Delta u$ for $u \in \mathrm{H}^2(\mathbb{R}^d)$ and, therefore, that

$$u_t = \Delta u = -\nabla E(u),$$

is indeed a gradient flow (in an infinite-dimensional space). Gradient flows in infinite-dimensional vector spaces (or even in metric spaces) are long standing (more recent)

---

7 This is justified for integrable and smooth functions simply by calculus knowledge. The validity extends to the more general case considered here by density.

fruitful tools in mathematics with applications in differential geometry and many areas of applied mathematics. The study of general functionals on infinite-dimensional spaces goes by the name of calculus of variations. The study of convex functionals falls under the label of convex analysis and allows generalizations to nondifferentiable functionals. Their powerful techniques find application in many areas of mathematics, as well.

9.3.2 The spectrum of a linear operator

In Section 8.2, we considered the problem of finding the eigenvalues of a square (symmetric/self-adjoint) matrix $A \in \mathbb{F}^{m\times m}$ for $\mathbb{F} = \mathbb{R}, \mathbb{C}$, and $m \in \mathbb{N}$. An eigenvalue $\lambda \in \mathbb{C}$, in this case, is characterized by the lack of injectivity of the map $A - \lambda$, i. e., the existence of $0 \neq x \in \mathbb{R}^m$ with $(A - \lambda)x = 0$. Eigenvalues λ_k and eigenvectors x_k of a symmetric matrix, counted according to their multiplicity, yield a particularly simple representation of A in the form

$$A = \sum_{k=1}^{m} \lambda_k x_k \overline{x}_k^\top.$$

This representation essentially tells us that in a basis of eigenvectors the linear map given by multiplication by the matrix A is represented by a diagonal matrix with the eigenvalues appearing on its diagonal. Studying the unbounded linear map,

$$-\Delta : \mathrm{H}^2_\pi \subset \mathrm{L}^2_\pi \to \mathrm{L}^2_\pi,$$

in a periodic context in Section 9.1 we derived a similar representation

$$-\Delta = \sum_{k\in\mathbb{Z}^d} |k|_2^2 \varphi_k \overline{\varphi}_k^\top,$$

maybe suggesting that (symmetric) linear maps between infinite-dimensional spaces can also be understood in terms of their eigenvalues and eigenfunctions. There are, however, significant differences between the finite and the infinite-dimensional cases. One such difference originates in the fact that $A \in \mathbb{R}^{m\times m}$ is one-to-one if and only if it is onto. Indeed, if A is not injective, there exists $0 \neq x \in \mathbb{R}^m$ such that $Ax = 0$, yielding a vanishing nontrivial linear combination $\sum_{k=1}^m x^k A_k^\bullet = 0$ of the columns of A. In this case, the m columns are not linearly independent and cannot span the image space $\mathbb{R}^m$, i. e., $R(A) \neq \mathbb{R}^m$. It follows that A is not surjective. Similarly, if A is not onto, then there must be a vanishing nontrivial linear combination of the m columns of A, showing that $N(A) \neq \{0\}$. This means that the set of eigenvalues $\sigma(A) \subset \mathbb{C}$ of a matrix $A \in \mathbb{R}^{m\times m}$ satisfies

$$\sigma(A) = \{\lambda \in \mathbb{C} \mid N(A - \lambda) \neq \{0\}\}$$

$$= \{\lambda \in \mathbb{C} \mid R(A - \lambda) \neq \mathbb{R}^m\}$$
$$= \{\lambda \in \mathbb{C} \mid A - \lambda \text{ is not invertible}\}.$$

This is no longer always true in infinite dimensions. We consider simple illustrative examples here as well as revisit $-\Delta$ on $\mathbb{R}^d$. We fix the space $\mathbb{R}_1^{\mathbb{N}}$ of sequences satisfying $\sum_{j\in\mathbb{N}} |x_j| < \infty$. Then the left-shift operator

$$S_l : \mathbb{R}_1^{\mathbb{N}} \to \mathbb{R}_1^{\mathbb{N}}, (x_j)_{j\in\mathbb{N}} = (x_1, x_2, \ldots) \mapsto (x_j)_{j\geqslant 2} = (x_2, x_3, \ldots)$$

is not injective since $S_l e_1 = 0$ but it is surjective. On the other hand, the right-shift operator

$$S_r : \mathbb{R}_1^{\mathbb{N}} \to \mathbb{R}_1^{\mathbb{N}}, (x_j)_{j\in\mathbb{N}} = (x_1, x_2, \ldots) \mapsto (0, x_1, x_2, \ldots)$$

is injective but not onto, since $e_1 \notin R(S_r)$. These are, however, not all the possible ways in which a linear map on an infinite-dimensional space can fail to be invertible. Consider the map

$$L : \mathbb{R}_1^{\mathbb{N}} \to \mathbb{R}_1^{\mathbb{N}}, \quad (x_j)_{j\in\mathbb{N}} \mapsto \left(\frac{x_j}{j}\right)_{j\in\mathbb{N}},$$

which is injective, has the inverse

$$M : R(L) \to \mathbb{R}_1^{\mathbb{N}}, \quad (x_j)_{j\in\mathbb{N}} \mapsto (jx_j)_{j\in\mathbb{N}},$$

which is not continuous, since it cannot hold that

$$|My|_1 \leqslant c\,|y|_1, \quad y \in R(L),$$

for any $c < \infty$. The latter can be seen by taking $(y^n)_{n\in\mathbb{N}} \in \mathbb{R}_1^{\mathbb{N}}$ given by

$$y^n = \left(1, \frac{1}{4}, \ldots, \frac{1}{n^2}, 0, 0, \ldots\right), \quad n \in \mathbb{N},$$

for which we have

$$|My^n|_1 = \sum_{j=1}^{n} \frac{1}{j} \to \infty \quad \text{as } n \to \infty,$$

while $|y^n|_1 \leqslant \frac{\pi^2}{6} < \infty$ for $n \in \mathbb{N}$. Notice that, since c_{00} is dense in $\mathbb{R}_1^{\mathbb{N}}$, so is $R(L)$ because $c_{00} \subset R(L)$. Since M is not continuous, it can, however, not be extended (continuously) to $\mathbb{R}_1^{\mathbb{N}}$.

Going back to $-\Delta$ on $\mathbb{R}^d$, observe that, for any $\lambda \in \mathbb{C}$, it holds that

$$-\Delta - \lambda = \mathcal{F}^{-1} M_{|\xi|^2 - \lambda} \mathcal{F},$$

where M_f denotes the linear operation of multiplication by the function f, i. e., $M_f(u) = fu$ for $u \in L^2(\mathbb{R}^d)$. In particular, if $\lambda \notin [0, \infty)$, we have that

$$(-\Delta - \lambda)^{-1} = \mathcal{F}^{-1} M_{\frac{1}{|\xi|^2 - \lambda}} \mathcal{F},$$

which, with the help of Plancherel's identity, entails that

$$\begin{aligned} \|(-\Delta - \lambda)^{-1} u\|_2 &= \left\| \frac{1}{|\xi|^2 - \lambda} \hat{u} \right\|_2 \leqslant \frac{1}{\operatorname{dist}(\lambda, [0, \infty))} \|\hat{u}\|_2 \\ &\leqslant c(\lambda) \|u\|_2, \quad u \in L^2(\mathbb{R}^d), \end{aligned}$$

and yields the invertibility of $-\Delta - \lambda$ for $\lambda \in \mathbb{C} \setminus [0, \infty)$. While the points $\lambda \in [0, \infty)$ are not eigenvalues in the sense that there is no $\varphi_\lambda \in L^2(\mathbb{R}^d)$ with $-\Delta u = \lambda u$, the latter equation does hold for $\varphi_\lambda^\pm(x) = e^{\pm i \xi \cdot x}$, $x \in \mathbb{R}^d$, with $\lambda = |\xi|_2^2$. These functions can be used to show[8] that

$$\inf_{\|u\|_2 = 1} \|(-\Delta - \lambda) u\|_2 = 0,$$

thus showing that $-\Delta - \lambda$ is not invertible for $\lambda \in [0, \infty)$. If it were, there would be a constant $c > 0$ with $\|(-\Delta - \lambda) u\|_2 \geqslant c \|u\|_2$ for $u \in L^2(\mathbb{R}^d)$ (why?). The functions $\varphi_\lambda^\pm$ are called approximate eigenfunctions, in this case. Infinite-dimensional spaces and linear maps between them are the subject of study of functional analysis, of which spectral theory is the branch that investigates the spectrum of linear operators. Complex analysis also plays a prominent role in the theory of operators. In order to illustrate this point, we conclude this section by briefly introducing and discussing the resolvent of a linear operator. The *resolvent set* of a linear operator A (bounded or unbounded) on a Banach space E is given by

$$\rho(A) = \{\lambda \in \mathbb{C} \mid A - \lambda \text{ is invertible}\},$$

and the *resolvent* of A is $(A - \lambda)^{-1}$ for $\lambda \in \rho(A)$. If $\lambda_0 \in \rho(A)$, then

$$\|(A - \lambda_0)^{-1}\|_{\mathcal{L}(E)} \leqslant c < \infty.$$

From this, we can infer that

$$\|\underbrace{(\lambda - \lambda_0)(A - \lambda_0)^{-1}}_{=B(\lambda)}\|_{\mathcal{L}(E)} < 1,$$

for $|\lambda_0 - \lambda| < \frac{1}{c}$. Then, for $\lambda \in \mathbb{B}_{\mathbb{C}}(\lambda_0, \frac{1}{c})$,

$$A - \lambda = A - \lambda_0 + \lambda_0 - \lambda = (A - \lambda_0)[\mathrm{id}_E - B(\lambda)]$$

8 Try to give a proof by approximating the functions $\varphi_\lambda^\pm$ by square integrable ones.

is invertible with inverse

$$(A-\lambda)^{-1} = \sum_{k=0}^{\infty} B(\lambda)^k (A-\lambda_0)^{-1} = \sum_{k=0}^{\infty} (\lambda-\lambda_0)^k (A-\lambda_0)^{-k-1},$$

since

$$\left[\mathrm{id}_E - B(\lambda)\right]^{-1} = \sum_{k=0}^{\infty} B(\lambda)^k,$$

for $\lambda \in \mathbb{B}_{\mathbb{C}}(\lambda_0, \frac{1}{c})$.[9] This shows that $\rho(A) \subset \mathbb{C}$ is open. Using the simple (resolvent) identity

$$\begin{aligned}(A-\mu)^{-1} - (A-\lambda)^{-1} &= (A-\mu)^{-1}[A-\lambda-(A-\mu)](A-\lambda)^{-1}\\ &= (\mu-\lambda)(A-\mu)^{-1}(A-\lambda)^{-1}, \quad \mu,\lambda \in \rho(A),\end{aligned}$$

we see that

$$\lim_{\mu\to\lambda} \frac{(A-\mu)^{-1} - (A-\lambda)^{-1}}{\mu-\lambda} = (A-\lambda)^{-2} \quad \text{in } \mathcal{L}(E),$$

which shows that

$$R_A : \rho(A) \to \mathcal{L}(E), \quad \lambda \mapsto (A-\lambda)^{-1},$$

is complex differentiable and that $\frac{d}{d\lambda}R_A(\lambda) = R_A^2(\lambda)$ for $\lambda \in \rho(A)$. This makes a whole range of techniques and results of complex analysis available in the study of linear operators and their spectral properties.

---

9 Show that this is indeed true and, in particular, that the series converges in the operator norm $\|\cdot\|_{\mathcal{L}(E)}$.

10 Probability

Due to our sometimes intrinsic, sometimes practical inability to understand or discover the precise laws that underlie natural phenomena, it can happen that we need to settle for the estimation of the likelihood of possible outcomes as opposed to fully explaining them. A commonly used example is that of coin tossing: while we know the physical laws that describe the motion of solid objects, the parameters involved in a coin toss are too many and not necessarily accurately measurable to make it possible to use a physical model to predict the outcome. In such cases, it is sometimes possible to obtain probabilistic statements instead. Our effort to understand the chances of events has led to the development of probability theory and statistics. Here, we give a very partial introduction to elementary probability making a connection with many topics that were covered so far in this book.

10.1 Axioms of probability

The intuitive understanding of probability we all share is that of *frequency of occurrence*. If we flip a coin $n \in \mathbb{N}$ times and collect the number $0 \leqslant n_H \leqslant n$ of heads observed, we would take the ratio

$$p = \frac{n_H}{n}$$

as some measure of the probability of observing heads. While we may not trust this measure if n is small, we tend to believe that a large number of flips would yield a good approximation of the “actual” probability. This requires, in particular, that we believe that there is an underlying probability of occurrence for events of interest. First and foremost, we need an *experiment* which can be repeated and which has well-defined *outcomes*. Rolling a die once is such an experiment, the outcomes of which are the face values $1, 2, \ldots, 6$. We therefore model an experiment by collecting its possible outcomes into a set S that we call the *sample space*. When rolling a die once, we have that

$$S = \{1, 2, 3, 4, 5, 6\}.$$

Events are collections of outcomes. If we care whether the outcome of a die roll is odd, for instance, we would be considering the event

$$\{1, 3, 5\} \subset S.$$

The probability of an outcome, or more in general of an event, would then be some number between 0 and 1 which aims at quantifying the “expected” frequency with which we observe the event of interest in a series of experiments. Probability therefore appears to be a function P that takes events, i. e., subsets $E \subset S$, as arguments. The set

https://doi.org/10.1515/9783110780925-010

S collects all possible outcomes. The value at an argument E is a number $P(E) \in [0,1]$. Since S contains all possible outcomes, we expect that $P(S) = 1$. This is almost all that is needed for a viable and useful definition of *probability*, which is formally defined as a function $P : \mathcal{E} \to [0,1]$ on a subset $\mathcal{E}$[1] $\subset 2^S$ of the power set

$$2^S = \{E \mid E \subset S\}$$

of a set S satisfying $P(S) = 1$ and the following *σ-additivity* condition:

$$P\Big(\bigcup_{k\in\mathbb{N}} E_k\Big) = \sum_{k=1}^{\infty} P(E_k),$$

which is required to hold for events $E_k \in \mathcal{E}$ that are pairwise disjoint, i. e., provided that

$$E_j \cap E_k = \emptyset \quad \text{for } j \neq k.$$

This latter condition is not as intuitive as the previous two and enforces a form of continuity of P. When the sample space $S = \{\omega_1, \omega_2, \omega_3, \dots\}$ is countable (or finite), all properties are automatically satisfied if

$$P(\{\omega_k\}) = p_k, \quad k \in \mathbb{N},$$

for $p_k \in [0,1]$ with $\sum_{k=1}^{\infty} p_k = 1$, $\mathcal{E} = 2^S$, and

$$P(E) = \sum_{\omega\in E} P(\{\omega\}) \quad \text{for } E \subset S.$$

If the sample space S is uncountable, then it can be shown (measure theory[2]) that there exist probabilities (in the above sense) that cannot be defined on the whole power set and one has to work with some $\mathcal{E} \subsetneq 2^S$.

If one flips a coin indefinitely and records the outcome $\omega_k \in \{H,T\}$ of each flip $k \in \mathbb{N}$, then the experiment has the sample space

$$S = \{H,T\}^{\mathbb{N}} = \{(\omega_k)_{k\in\mathbb{N}} \mid \omega_k \in \{H,T\} \text{ for } k \in \mathbb{N}\}.$$

If the coin is fair, then it holds that

$$P(\omega_j = H) = P(\{\omega_j = H\}) = .5$$

for any fixed $j \in \mathbb{N}$, where $\{w_j = H\}$ is shorthand for the event given by the set $\{(\omega_k)_{k\in\mathbb{N}} \mid w_j = H\} \subset 2^S$. If one is interested in the event $T_{\leqslant 4}$ that no heads are flipped

1 For the definition to make sense, $\mathcal{E}$ must be such that $S \in \mathcal{E}$, that $E^c \in \mathcal{E}$ whenever $E \in \mathcal{E}$, and that $\bigcup_{k\in N} E_k \in \mathcal{E}$ whenever $E_k \in \mathcal{E}$ for $k \in \mathbb{N}$. Such a $\mathcal{E}$ is called σ-algebra.

2 See, for instance, the textbook by Folland, *Real Analysis: Modern Techniques and Their Applications*.

in the first 4 flips, then

$$T_{\leqslant 4} = \{(\omega_k)_{k\in\mathbb{N}} \mid \omega_j = T \text{ for } j = 1,\ldots,4\} = \bigcap_{k=1}^{4}\{\omega_k = T\}.$$

In order to help the intuition, we observe that the continuity property of probability implies another condition which more explicitly resembles a sort of continuity. Take a sequence of $(E_k)_{k\in\mathbb{N}}$ of events in $\mathcal{E}$ with the property that, if

$$\text{either } E_k \subset E_{k+1}\ \forall k \in \mathbb{N} \quad \text{or} \quad E_k \supset E_{k+1}\ \forall k \in \mathbb{N},$$

i. e., if the sequence is "increasing" or it is "decreasing", then it holds that

$$P\Big(\lim_{k\to\infty} E_k\Big) = \lim_{k\to\infty} P(E_k)$$

if we set

$$\lim_{n\to\infty} E_n = \bigcup_{k\in\mathbb{N}} E_k \quad \text{or} \quad \lim_{n\to\infty} E_n = \bigcap_{k\in\mathbb{N}} E_k,$$

respectively. To see this in the case of an increasing sequence of events, we can keep $E_1 = F_1$ and replace E_k by $F_k = E_k \setminus E_{k-1}$ for $n \geqslant 2$ to obtain pairwise disjoint sets for which $\bigcup_{k\in\mathbb{N}} E_k = \bigcup_{k\in\mathbb{N}} F_k$ and argue based on

$$P\Big(\bigcup_{k\in\mathbb{N}} E_k\Big) = P\Big(\bigcup_{k\in\mathbb{N}} F_k\Big) = \sum_{k=1}^{\infty} P(F_k) = \lim_{n\to\infty}\sum_{k=1}^{n} P(F_k) = \lim_{n\to\infty} P(E_n)$$

that the defining condition implies the identity for increasing sequences of events. The argument for decreasing sequences is similar and left as an exercise. If P is additive, i. e., if it satisfies

$$P\Big(\bigcup_{k=1}^{n} E_k\Big) = \sum_{k=1}^{n} P(E_k) \quad \text{provided } E_j \cap E_k = \emptyset \text{ for } j \neq k,$$

for any $n \in \mathbb{N}$, then σ-additivity and the above continuity condition are equivalent (check this (Q1)).

10.2 Conditional probability and independence

One the most important concepts, if not the most important concept, in probability theory is that of *conditional probability*, which is closely related to that of independence. Given a probability space $(S, \mathcal{E}, P)$, fix an event $F \in \mathcal{E}$ with $P(F) > 0$ and define the *conditional probability* $P(E|F)$ *of* E *given* F by the expression

$$P(E|F) = \frac{P(E\cap F)}{P(F)}, \quad E \in \mathcal{E}.$$

Thinking of probability as frequency of occurrence, this definition is quite intuitive in the sense that it measures the frequency of concurrent occurrence of E and F among the occurrences of F. It can be verified (do it for yourself) that

$$P(\cdot|F) : \mathcal{E} \to [0,1], \quad E \mapsto P(E|F),$$

is itself a probability on S. Two events $E, F \in \mathcal{E}$ are *independent* with respect to the probability P iff

$$P(E|F) = P(E) \quad \text{or, equivalently, if} \quad P(F|E) = P(F),$$

i. e., iff the probability of E is not altered by the knowledge of F occurring or, equivalently, iff that of F is not changed by knowing that E has occurred. The symmetry is apparent in the equivalent condition

$$P(E \cap F) = P(E)P(F).$$

Conditional probability allows one to consider a series of circumstances and relate the total probability of an event to its probability under the given circumstances. Letting $F \in \mathcal{E}$ be a circumstance, so that F^c describes its nonoccurrence, we have that, for any event $E \in \mathcal{E}$, the so-called *Bayes' formula*

$$\begin{aligned} P(E) &= P((E \cap F) \cup (E \cap F^c)) \doteq P(E \cap F) + P(E \cap F^c) \\ &= P(E|F)P(F) + P(E|F^c)P(F^c), \end{aligned}$$

holds. It has the nice interpretation that the probability of E can be computed as its probability given the circumstance F times the frequency of F added to the probability of E given F^c times the frequency of F^c. Convince yourself that the same remains true for any pairwise disjoint "set" of circumstances $F_j \in \mathcal{E}$ that partitions the sample space S, i. e., satisfying $\bigcup_{j\in\mathbb{N}} F_j = S$, i. e., that

$$P(E) = \sum_{j\in\mathbb{N}} P(E|F_j)P(F_j)$$

Independence is mainly used to model situations where an experiment is repeated with the intuitive understanding that the outcome of a trial does not influence that of another. When repeatedly flipping a fair coin, it is reasonable to assume that the outcome of a flip has no influence on the outcome of any other additional flips (unless the flips are performed by a very precise machine with the exact same settings each time). Then, when computing probabilities, independence yields useful simplifications. Take for instance $T_{\leqslant 4}$ for which we have that[3]

$$P(T_{\leqslant 4}) = P(\omega_1 = \cdots = \omega_4 = T) = P\Big(\bigcap_{k=1}^{4}\{\omega_k = T\}\Big)$$
$$= \prod_{k=1}^{4} P(\omega_k = T) = \frac{1}{2^4} = \frac{1}{16}.$$

10.3 Random variables

Another central concept of probability theory is that of *random variable*. Before giving a formal definition, we try and motivate it by considering a concrete example and comparing it to that of a deterministic variable. Let x be the length of a person's foot, say yours. You can consider it as the independent variable x of the function size that returns the corresponding shoe size size(x). You measure your foot and determine your shoe size, and like you, anybody has a determined foot length and a corresponding shoe size. Consider now a shoe factory and the basic question of how many shoes of each given size to produce and distribute. Foot length is no longer a deterministic variable since the factory does not know all its customers, it does typically not even sell directly to the public! Even if it did, it would not be in a position to know who will want to buy a pair of shoes at any given time. Thus the best the factory can do is consider foot length as a random variable X, which it does not know the value of. It does know what the possible values are: any number between the smallest and largest foot lengths $L_{\min}$ and $L_{\max}$ ever observed, i. e., $X \in [L_{\min}, L_{\max}]$. It can also try to measure the prevalence or frequency of certain foot sizes by simple statistical analysis of the target population and/or of past sales. With this information, it can determine the probability $P(X = L)$ of each foot length $L \in [L_{\min}, L_{\max}]$ or, more realistically the probability $P(L_i \leqslant X < L_{i+1})$ of the foot length to fall in certain intervals $[L_i, L_{i+1})$ $(i = 0, \dots, n-1)$, where $L_{\min} = L_0 < L_1 < \cdots < L_n = L_{\max}$, maybe corresponding to shoe size. In this case, X cannot be thought of as an independent variable because its value does depend on the specific buyer, which the factory cannot know at the time of production. This motivates the formal definition of *random variable with values in* T as a function $X : S \to T$ defined on a probability space $(S, \mathcal{E}, P)$ and taking values in a set T. It is then possible to obtain information about the probability of certain values by considering

$$P(X \in U) = P(\{X \in U\}) = P(\{\omega \in S \mid X(\omega) \in U\}),$$

3 Notice that independence of multiple sets $E_1, \dots, E_n \in \mathcal{E}$ amounts to the validity of

$$P\Big(\bigcap_{j=1}^{k} E_{n_j}\Big) = \prod_{j=1}^{k} P(E_{n_j}),$$

for any choice of $1 \leqslant n_1 < n_2 < \cdots < n_k \leqslant n$ and of $2 \leqslant k \leqslant n$. This is more than mere pairwise independence.

for $U \subset T$ such that $X^{-1}(U) \in \mathcal{E}$.[4] Notice that, when computing the shoe size, we would consider the composition size $\circ X$ = size(X). Thus we can consider a function of a random variable just as we can consider a function of a deterministic variable. Notice that size $\circ X$ is itself a random variable, albeit with values in another set. Restricting our attention to real valued random variables ($T \subset \mathbb{R}$), we distinguish two cases: *discrete and continuous* random variables. A discrete random variable X is characterized by having a discrete range

$$X(S) = \{x_1, x_2, \dots\} \subset \mathbb{R}.$$

In this case, the random variable is fully understood in terms of

$$p_X(x_k) = P(X = x_k) = p_k \in (0, 1], \quad k \in \mathbb{N},$$

where we can assume the positivity of p_k since otherwise x_k would actually not be a value of X in the first place. Indeed, we have that

$$P(X \in U) = \sum_{x \in U} p_X(x) \quad \text{for any } U \subset \mathbb{R}.$$

Notice that the sum/series always makes sense since $S = X^{-1}(\mathbb{R})$, $P(S) = 1$, and hence

$$0 \leqslant P(X \in U) \leqslant \sum_{k \in \mathbb{N}} p_k = 1 \quad \text{for any } U \subset \mathbb{R}.$$

The function $p_X : \mathbb{R} \to [0, 1], x \mapsto p(X = x)$ is known as the *probability mass function* of the discrete random variable X. A *continuous random variable* X does not assign a positive probability to any specific value but is characterized by the existence of a so-called *probability density function* $f_X : \mathbb{R} \to [0, \infty)$ for which it holds that

$$P(X \in U) = \int_U f(x)\, dx \quad \text{for } U \subset \mathbb{R},$$

where f_X is assumed to be piecewise continuous.[5] It is enough to think of U as an interval $[a, b]$ for $-\infty \leqslant a \leqslant b \leqslant \infty$ in which case we have that

$$P(X \in [a, b]) = \int_a^b f(x)\, dx.$$

Take a minute to consider the parallels between the concepts of mass and density in physics and the corresponding concepts of probability mass and probabil-

4 The requirement that $\{X \in U\} \in \mathcal{E}$ is necessary whenever $\mathcal{E} \neq 2^S$ and, in a full treatment of the subject, leads to the definition of *measurability* of X.

5 It would actually be enough to assume that $f \in L^1(\mathbb{R}, [0, \infty))$ here, i. e., Lebesgue integrable in the sense of measure theory.

ity density functions. Show (Q2) that, for any random variable with real values, at most a countable number of the values can have positive probability, i. e., that the set $\{x \in \mathbb{R} \,|\, P(X = x) > 0\}$ is at most countably infinite (consider the sets $\{x \in \mathbb{R} \,|\, P(X = x) \geqslant \frac{1}{n}\}$ for $n \in \mathbb{N}$ and their relation to it).

Given any random variable X with real values, we can obtain insight into how the values are distributed (in terms of their probability) by looking at the function $F_X : \mathbb{R} \to \mathbb{R}$ defined by

$$F_X(x) = P(X \leqslant x), \quad x \in \mathbb{R}.$$

It is called *cumulative distribution function*. Given any random variable X, convince (Q3) yourself that:

(i) It holds that $\lim_{x \to -\infty} F_X(x) = 0$.
(ii) It holds that $\lim_{x \to \infty} F_X(x) = 1$.
(iii) F_X is nondecreasing, i. e., $F_X(x) \leqslant F_X(\tilde{x})$, whenever $x \leqslant \tilde{x}$.
(iv) F_X is right continuous and admits left limits in the sense that

$$\lim_{\tilde{x} \searrow x} F_X(\tilde{x}) = F(x) \quad \text{and} \quad \lim_{\tilde{x} \nearrow x} F_X(\tilde{x}) \text{ exists},$$

for any $x \in \mathbb{R}$.

Draw a generic graph of the cumulative distribution function of a discrete random variable and one of a continuous random variable. Given a random variable X with real values, we obtain a probability measure on $\mathbb{R}$, if we define

$$P_X(U) = \sum_{x \in U} p_X(x), \quad U \subset \mathbb{R},$$

or

$$P_X(U) = \int_U f_X(x)\,dx, \quad U \in \mathcal{E},$$

for a discrete or continuous random variable X, respectively. The domain of definition $\mathcal{E}$ is intentionally left vague but contains all intervals.

The simplest example of a discrete random variable is one with a single value (which makes it deterministic, but we can still consider it a random variable). If that value is 0, $X \equiv 0$, then its cumulative distribution function is given by

$$h_X(x) = \begin{cases} 0, & x < 0, \\ 1, & x \geqslant 0 \end{cases},$$

while its probability mass function is the function p_X, which vanishes everywhere except at $x = 0$, where $p_X(0) = 1$. It is often thought as having probability density function f_X concentrated in 0. While we will not make it rigorous here, it can be shown that

$$f_X = \delta_0,$$

where δ_0 is the probability measure on $\mathbb{R}$ with domain $\mathcal{E} = 2^{\mathbb{R}}$ defined by

$$\delta_0(U) = \begin{cases} 1, & 0 \in U, \\ 0, & 0 \notin U. \end{cases}$$

In particular, f_X is not a function. Making this rigorous would require measure theory, so that without it, it is more convenient to talk about probability mass function for discrete random variables, thus avoiding the discussion of *singular* density functions like δ_0 altogether.

Another example of a discrete random variable is the total number X of heads observed after $n \in \mathbb{N}$ tosses of a coin, for which the probability of heads is $p \in (0,1)$. Clearly, the coin is fair only when $p = .5$. This random variable has $1, 2, \dots, n$ as its values and it holds that

$$P(X = k) = \binom{n}{k} p^k (1-p)^{n-k}, \quad k = 1, \dots, n.$$

When flipping a coin indefinitely, the first toss resulting in heads is also a discrete random variable X for which $X(S) = \mathbb{N}$ and

$$P(X = k) = (1-p)^{k-1} p, \quad k \in \mathbb{N}.$$

Notice that (as expected)

$$\sum_{k=1}^{\infty} P(X = k) = p \sum_{k=1}^{\infty} (1-p)^{k-1} = \frac{p}{1-(1-p)} = 1.$$

If we randomly pick a real number in the interval $[-1, 1]$ according to the rule that each number should be just as likely to be chosen as any other and denote by X its value, then X is a continuous random variable with $X(S) = [-1, 1]$ and with

$$f_X(x) = \begin{cases} \frac{1}{2}, & x \in [-1, 1], \\ 0, & x \notin [-1, 1]. \end{cases}$$

In this case, we say that X is *uniformly distributed* in $[-1, 1]$ and that X is a uniform random variable. Notice that

$$\int_{-\infty}^{\infty} f_X(x)\, dx = \int_{-1}^{1} \frac{1}{2}\, dx = 1,$$

and that

$$F_X(x) = \begin{cases} 0, & x < -1, \\ \frac{x+1}{2}, & x \in [-1, 1], \\ 1, & x > 1. \end{cases}$$

We conclude this section by establishing a connection between the cumulative distribution and probability density functions of a continuous random variable. Take $x \in \mathbb{R}$ and $h > 0$ ($h < 0$ can be handled similarly with the necessary modifications). Then

$$\begin{aligned} \int_x^{x+h} f_X(y)\,dy &= P(X \in [x, x+h]) = P(x \leqslant X \leqslant x+h) \\ &= P(X \leqslant x+h) - P(X \leqslant x) \\ &= F_X(x+h) - F_X(x), \end{aligned}$$

since $\{X \leqslant x+h\} = \{X \leqslant x\} \cup \{x < X \leqslant x+h\}$ with disjoint union. Dividing both sides of this identity by h, taking the limit as $h \searrow 0$, and using the fundamental theorem of calculus, we see that

$$f_X(x) = F_X'(x), \qquad x \in \mathbb{R},$$

at least at points where f_X is continuous.

10.4 A discrete random walk and the diffusion equation

Next, we consider the following process which takes places in a two-dimensional roster as depicted below.

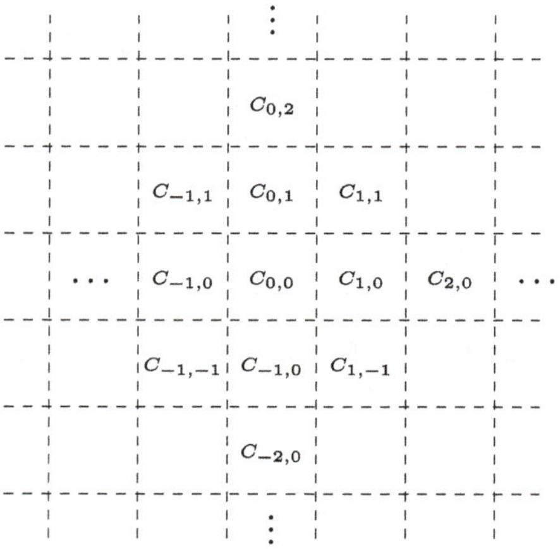

Each square cell $C_{i,j}$ has side length $dx > 0$ and is centered at the point with coordinates $(idx, jdx) \in \mathbb{R}^2$. We assume that a particle is initially, i. e., at time $t = 0$, found in cell $C_{0,0}$ with probability 1. We denote by X^n the location of the particle at time $t = ndt$ for a time step $dt > 0$. For each $n \in \mathbb{N}$, X^n is a random variable and

$$P(X^0 \in C_{i,j}) = \begin{cases} 1, & (i,j) = (0,0), \\ 0, & (i,j) \neq (0,0). \end{cases}$$

We do not care where exactly the particle is in a cell and, therefore, assume that X^0 is uniformly distributed in $C_{0,0}$, i. e., that

$$f_{X^0}(y) = \begin{cases} \frac{1}{dx^2}, & y \in C_{0,0}, \\ 0, & y \notin C_{0,0}. \end{cases}$$

We assume that, in each time step, the particle moves anywhere in a cell that is to the left, to the right, below, or above the current cell, each with the probability 1/4. Then, for instance, we would have that

$$P(X^1 \in C_{i,j}) = \begin{cases} \frac{1}{4}, & (i,j) \in \{(-1,0),(1,0),(0,-1),(0,1)\} \\ 0, & (i,j) \notin \{(-1,0),(1,0),(0,-1),(0,1)\}, \end{cases}$$

and that

$$f_{X^1}(y) = \begin{cases} \frac{1}{4dx^2}, & y \in C_{-1,0} \cup C_{1,0} \cup C_{0,-1} \cup C_{0,1}, \\ 0, & y \notin C_{-1,0} \cup C_{1,0} \cup C_{0,-1} \cup C_{0,1}, \end{cases}$$

again assuming that we do not record the exact location of the particle in a cell but rather think of it as being anywhere in the cell with equal probability. As the process evolves step-by-step, we are interested in what can be said about the probability of finding the particle in any given cell at any given step. Let us therefore define $p^n_{i,j}$ as the probability density of finding the particle in cell $C_{i,j}$ in step n, i. e., at time ndt. Then

$$P(X^n \in C_{i,j}) = \int_{C_{i,j}} f_{X^n}(y)\,dy = p^n_{i,j} dx^2 \quad \text{for } n \in \mathbb{N} \text{ and } (i,j) \in \mathbb{Z}^2.$$

Since the cells partition the whole plane, we can use Bayes' formula to see that

$$\begin{aligned} p^{n+1}_{i,j} dx^2 &= P(X^{n+1} \in C_{i,j}) = \sum_{k,l \in \mathbb{Z}} P(X^{n+1} \in C_{i,j} | X^n \in C_{k,l}) P(X^n \in C_{k,l}) \\ &= \frac{dx^2}{4} [p^n_{i-1,j} + p^n_{i+1,j} + p^n_{i,j-1} + p^n_{i,j+1}] \quad \text{for } (i,j) \in \mathbb{Z}^2,\ n \geqslant 0, \end{aligned}$$

since a particle found in cell $C_{i,j}$ in step $n+1$ must have been found in any of the four adjacent cells in the previous step with equal probability. Upon subtracting $p^n_{i,j}$ from both sides of the identity and dividing by dt, we arrive at

$$\frac{p^{n+1}_{i,j} - p^n_{i,j}}{dt} = \frac{dx^2}{4dt} \frac{p^n_{i-1,j} + p^n_{i+1,j} + p^n_{i,j-1} + p^n_{i,j+1} - 4p^n_{i,j}}{dx^2},$$

for $(i,j) \in \mathbb{Z}^2$, $m \geqslant 0$. If we think of $p^n_{i,j}$ as $p(ndt, idx, jdx)$ for some underlying function p (that we assume exists) where $t = ndt$ and $x = (i,j)dx$, then

$$\frac{p^{n+1}_{i,j} - p^n_{i,j}}{dt} = \frac{p(t+dt,x) - p(t,x)}{dt} \simeq p_t(t,x) \quad \text{for } dt \simeq 0,$$

and

$$\begin{aligned}
\frac{p^n_{i-1,j} + p^n_{i+1,j} - 2p^n_{i,j}}{dx^2} &= \frac{p(t,x-dxe_1) + p(t,x+dxe_1) - 2p(t,x)}{dx^2} \\
&\simeq \partial_1^2 p(t,x) \quad \text{for } dx \simeq 0, \\
\frac{p^n_{i,j-1} + p^n_{i,j+1} - 2p^n_{i,j}}{dx^2} &= \frac{p(t,x-dxe_2) + p(t,x+dxe_2) - 2p(t,x)}{dx^2} \\
&\simeq \partial_2^2 p(t,x) \quad \text{for } dx \simeq 0.
\end{aligned}$$

Thus, given $(t,x) \in (0,\infty) \times \mathbb{R}^2$, if we let $dt, dx \to 0$ in such a way that

$$\frac{dx^2}{4dt} = D > 0, \quad ndt \to t \text{ and } (idx, jdx) \to x,$$

we may guess that the limiting probability density $p : (0,\infty) \times \mathbb{R}^2 \to [0,\infty)$ (if it exists) satisfies

$$p_t(t,\cdot) = D\Delta p(t,\cdot) \quad \text{for } t > 0,\ p(0,\cdot) = \delta_0.$$

As for the initial condition, notice, somewhat formally, that

$$\int \varphi(y) f_{X^0}(y)\, dy \longrightarrow \varphi(0) = \text{“} \int \varphi(y)\delta_0(y)\, dy \text{”}$$

for any smooth function $\varphi : \mathbb{R}^2 \to \mathbb{R}$. This can be interpreted as saying that the limiting probability density is singular and concentrated in 0, i. e., that

$$f_{X_0} \to \delta_0 \quad \text{as } dx \to 0.$$

We conclude that, in the limit of dt and dx small, the probability density function $p^n_{i,j}$ is approximated by $p(t,x)$, where p is the solution of the diffusion equation provided that $t = ndt$ and $x = dx(i,j)$. We know that

$$p(t,x) = \frac{1}{2\pi Dt} e^{-|x|_2^2/4Dt}, \quad (t,x) \in (0,\infty) \times \mathbb{R}^2,$$

so that we finally obtain that

$$P(X^n \in U) \simeq \frac{1}{2\pi Dt} \int_U e^{-|x|_2^2/4Dt}\, dx,$$

for $U \in \mathcal{E}$, where $\mathcal{E}$ includes open and closed sets. It can be shown, but this requires significant additional technical developments, that there is a so-called stochastic process $\{X^t \mid t \geqslant 0\}$ comprised of random variables X^t with

$$f_{X^t}(x) = \frac{1}{2\pi t} e^{-|x|_2^2/4t}, \quad x \in \mathbb{R}^2 \text{ for } t > 0,$$

and $f_{X^0} = \delta_0$ and satisfying additional properties. It is called (2-dimensional) Brownian motion.

10.5 Expected value of a random variable

We begin with a motivational example. Assume that we are given a data set $d_1, \dots, d_n$ and would like to determine its mean/average μ. We would likely automatically compute

$$\mu = \frac{d_1 + \dots + d_n}{n} = \frac{1}{n} \sum_{k=1}^{n} d_k.$$

If the observed values are all in the set $\{1, 2, \dots 6\}$, for instance, there is an arguably more efficient way to do this, which is obtained by first determining the number of times n_i each value $i = 1, \dots, 6$ occurs in the data set and then computing

$$\mu = \frac{1}{n} \sum_{i=1}^{6} i\, n_i = \sum_{i=1}^{6} i \frac{n_i}{n}.$$

The latter sum can be interpreted as a weighted average of the possible values, where the weights are given by the values' relative frequency of occurrence $\frac{n_i}{n} \in [0,1]$ in the data set. It clearly holds $\sum_{i=1}^{6} \frac{n_i}{n} = 1$. If we recall the intuitive understanding of probability as frequency of occurrence, we can define the *expected value* (mean/average) of a discrete random variable X with values $\{x_1, x_2, \dots\} \subset \mathbb{R}$ by

$$E[X] = \sum_{i \in \mathbb{N}} x_i p_X(x_i) = \sum_{i \in \mathbb{N}} x_i P(X = x_i).$$

For a concrete situation, roll a die n times and denote the outcomes of the rolls by $d_1, \dots, d_n$, where clearly $d_i \in \{1, 2, \dots, 6\}$. The average value rolled is given by

$$\frac{1}{n}\sum_{i=1}^{n} d_i = \sum_{i=1}^{n} i\frac{n_i}{n},$$

where $n_i = |N_i|$ is the number of elements in the set

$$N_i = \{k \in \{1, 2, \dots, n\} \mid d_k = i\},$$

for $i = 1, \dots, 6$. If the die is fair, we expect that $\frac{n_i}{n} \simeq \frac{1}{6}$, at least if n is large. This is reflected in

$$E[X] = \sum_{i=1}^{6} i\frac{1}{6} = \sum_{i=1}^{6} i\,P(X = i).$$

if X denotes the outcome of a roll and explains the definition.

10.6 The gambler's ruin problem

Consider a game with two players (A and B), which consists in repeatedly flipping a coin that shows heads with probability $p \in (0, 1)$. Each time *heads* comes up, player B gives a chip to player A, else player B receives one from A. The game ends when one of the players runs out of chips. If A starts with m chips of a total of N chips in the game (i. e., B begins with $N - m$ chips), what is the probability of A winning the game? A possible unfolding of a fair game ($p = .5$) is depicted below, in which player A eventually loses starting with 20 of a total of 40 chips.

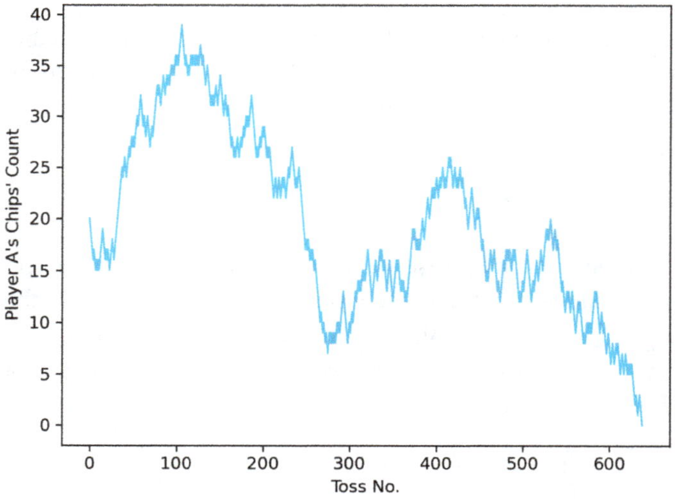

10.6.1 Winning probability

If W_m is the event of winning the game with m chips initially and H/T the events of flipping heads/tails in the first toss, Bayes' formula yields that

$$\begin{aligned} w_m &= P(W_m) = P(W_m|H)P(H) + P(W_m|T)P(T) \\ &= pP(W_{m+1}) + (1-p)P(W_{m-1}) = p\,w_{m+1} + (1-p)\,w_{m-1}, \end{aligned}$$

if $w_k = P(W_k)$ for $k = 0, \ldots, N$ and $p \in (0,1)$ is the probability of flipping heads. Notice that $w_0 = 0$ and $w_N = 1$ since A cannot play, let alone win, with no initial chips, while victory is assured starting with all the chips. This information is enough to recover the actual values of w_m. Indeed, it follows from the above that

$$\begin{aligned} w_{m+1} - w_m &= \frac{1-p}{p}(w_m - w_{m-1}) = \cdots = \left(\frac{1-p}{p}\right)^m (w_1 - w_0) \\ &= r^m w_1, \quad m = 1, \ldots, N-1, \end{aligned}$$

for $r = \frac{1-p}{p}$. This entails that

$$w_m = \sum_{k=0}^{m-1} (w_{m+1} - w_m) = w_1 \sum_{k=0}^{m-1} r^m = \begin{cases} \frac{1-r^m}{1-r} w_1, & r \neq 1, \\ m w_1, & r = 1. \end{cases}$$

Notice that $r = 1$ iff $p = \frac{1}{2}$, i. e., iff the coin is fair. Then we use

$$1 = w_N = \begin{cases} \frac{1-r^N}{1-r} w_1, & r \neq 1, \\ N w_1, & r = 1, \end{cases}$$

to find the value of w_1, which finally gives

$$w_m = \begin{cases} \frac{1-r^m}{1-r^N}, & r \neq 1, \\ \frac{m}{N}, & r = 1. \end{cases}$$

It follows that a gambler with limited (finite) resources ($m < \infty$) playing a fair game with an opponent of infinite wealth ($N = \infty$), such as a casino, is bound to eventually go broke ($w_m = 0$), i. e., meet her ruin.

In the special case when $p = \frac{1}{2}$ and N is large, it is possible to interpret the system

$$\begin{cases} w_m = \frac{1}{2} w_{m+1} + \frac{1}{2} w_{m-1}, & m = 1, \ldots, N-1, \\ w_m = \frac{m}{N}, & m = 0, N. \end{cases}$$

as the discrete manifestation of a boundary value problem. In order to see this, we think of the fraction $\frac{m}{N} \in [0,1]$ of chips in player A's possession and of the corresponding winning probability w_m as the value $w(\frac{m}{N})$ of an unknown function w defined on $[0,1]$ with values in the same interval. As it holds formally that

$$\begin{aligned} w_{xx}(x) &\simeq \frac{1}{h^2}(w(x+h) - 2w(x) + w(x-h)) \\ &= N^2(w_{m+1} - 2w_m + w_{m-1}) = 0 \end{aligned}$$

for $x = \frac{m}{N} \in (0,1)$ and $h = \frac{1}{N}$, we expect the validity of

$$\begin{cases} w_{xx} = 0, & x \in (0,1) \\ w(x) = x, & x = 0,1 \end{cases},$$

in the limit as $N \to \infty$. It is easily seen that the solution of this boundary value problem is given by $w(x) = x$ for $x \in [0,1]$. This, not only approximates the formula we derived for the corresponding probability, but actually coincides with it

$$w_m = \frac{m}{N} = w\left(\frac{m}{N}\right).$$

10.6.2 A pool of games

Before turning our attention to another interesting question about the gambler's ruin problem, we look at the winning probability problem from a slightly different viewpoint. Instead of assuming that player A starts the game with m chips, we think of many people, maybe in a casino, who are, in pairs, simultaneously playing the game. In each pair, one of the players is designated as player A. Each player starts with a different initial amount of chips. In order to follow all games simultaneously, we track the fraction π_m of players A who start with $m \in \{0,1,\dots,N\}$ chips. We keep the total number of chips in each game fixed at N. We use the vector

$$\pi = \begin{bmatrix} \pi_0 & \pi_1 & \dots & \pi_N \end{bmatrix}$$

to encode the state of the whole pool. Notice that $\sum_{j=0}^{N} \pi_j = 1$. Then we can ask about the time evolution of a pool that starts in a given state π^0, where time corresponds to number of flips. We can think of π_j^k as the probability $P(C_A^k = j)$ of finding a randomly chosen game with player A holding j chips after k coin flips. Here, we denote the number of chips in player A's possession after k flips by C_A^k. Using conditional probability and Bayes' formula, we see that

$$\pi_j^{k+1} = P(C_A^{k+1} = j) = \sum_{l=0}^{N} P(C_A^{k+1} = j | C_A^k = l) P(C_A^k = l)$$

$$
\begin{aligned}
&= P(C_A^{k+1} = j | C_A^k = j-1) P(C_A^k = j-1) \\
&\quad + P(C_A^{k+1} = j | C_A^k = j+1) P(C_A^k = j+1) \\
&= p\pi_{j-1}^k + (1-p)\pi_{j+1}^k,
\end{aligned}
$$

since any player A will gain one chip with probability p and lose one with probability $1-p$. We read this equation with the additional understanding that $\pi_{-1}^k = 0 = \pi_{N+1}^k$ in which case it is valid for all $j = 0, \dots, N$. We see that one step in the game causes a change in the probabilities, which can be captured by vector-matrix multiplication

$$\pi^k = \pi^{k-1} M = \cdots = \pi^0 M^k, \quad k \geqslant 1,$$

where the rows of the matrix $M \in [0,1]^{(N+1)\times(N+1)}$ are given by

$$
\begin{aligned}
M_\bullet^0 &= [1 \quad 0 \quad \dots \quad 0] = e_1^\top, \\
M_\bullet^N &= [0 \quad 0 \quad \dots \quad 1] = e_{N+1}^\top, \\
M_\bullet^i &= [0 \quad \dots \quad 0 \quad (1-p) \quad 0 \quad p \quad 0 \quad \dots \quad 0] = (1-p)e_{i-1}^\top + pe_{i+1}^\top,
\end{aligned}
$$

and the last identity holds for $i = 1, \dots, N-1$. The entries of this matrix can be interpreted as transition probabilities: M_j^i gives the probability of going from having i chips in one step to having j chips in the next step. The first row therefore indicates that you will still have no chips in the next step, if you have none currently. The last column reflects the fact that you will still have all chips in the next step, if you possess them all at present. All other rows capture the fact that the numbers of chips can only change by one in a step with probabilities as given. Notice that the row vectors $e_1^\top$ and $e_{N+1}^\top$ are left eigenvectors[6] of M to the eigenvalue 1. Indeed,

$$vM = v \quad \text{for } v = e_1^\top, e_{N+1}^\top.$$

This means that a pool where all players A have no chips does no longer evolve (all games are over). The same is true if all players A own all chips in the game. Not only can we interpret vector-matrix multiplication probabilistically, but also matrix-matrix multiplication. Let us, for instance, take M^2 and find an interpretation of its entries. Bayes' formula[7] gives

6 Left eigenvectors v of M are the transpose of regular (right) eigenvectors $v^\top$ of the transposed matrix $M^\top$.

7 Here, we apply Bayes' formula not to the probability P, but rather to the conditional probability $P(\cdot|F)$. Its validity follows from the fact that $P(\cdot|F)$ is itself a probability for any fixed F (check this).

$$\begin{aligned}P(C_A^{k+2} = l|C_A^k = i) &= \sum_{j=0}^{N} P(C_A^{k+2} = l|C_A^{k+1} = j \,\&\, C_A^k = i)P(C_A^{k+1} = j|C_A^k = i)\\ &= \sum_{j=0}^{N} M_l^j M_j^i = (M^2)_l^i,\end{aligned}$$

where we used the fact that

$$P(C_A^{k+2} = l|C_A^{k+1} = j \,\&\, C_A^k = i) = P(C_A^{k+2} = l|C_A^{k+1} = j),$$

since we are assuming that the coin flips are independent and the state of the pool at any step only depends on its state in the previous step. Independence of the coin flips also yields the validity of the two above formulæ for any $k = 0, 1 \ldots$ This argument can be applied inductively to see that $(M^n)_j^i$ amounts to the probability of transitioning from having i chips to having j in n steps. The probability w_m of ending a game with a win starting with m chips initially is therefore given by

$$w_m = \lim_{n\to\infty} (M^n)_N^m,$$

whereas that of losing, denoted by l_m, is correspondingly given by

$$l_m = \lim_{n\to\infty} (M^n)_0^m.$$

The limit needs to be taken since victory or defeat can intervene after any number of coin flips. It can be verified that all rows of M^n add to 1 for any $n \in \mathbb{N}$ so that this remains valid for the limiting matrix $M^\infty = \lim_{n\to\infty} M^n$. Next, we show that $(M^\infty)_j^i = 0$ for each $i \neq 0, N$ and $j \neq 0, N$, i. e., the only two nonzero columns of M^∞ are the first and the last. The reader is asked to verify that $(M^\infty)_j^0 = 0$ for $j = 1, \ldots, N$ and $(M^\infty)_j^N = 0$ for $j = 0, \ldots, N-1$. As for the other entries, we take an arbitrary

$$\pi = \begin{bmatrix} \pi_0 & \pi_1 & \ldots & \pi_N \end{bmatrix} \in \mathbb{R}^{1\times(N+1)}$$

and observe that

$$(\pi M)_0 = \pi_0 + (1-p)\pi_1, \quad (\pi M)_1 = (1-p)\pi_2, \tag{10.1}$$

$$(\pi M)_j = p\pi_{j-1} + (1-p)\pi_{j+1} \quad \text{for } j = 2, \ldots, N-2, \tag{10.2}$$

$$(\pi M)_{N-1} = p\pi_{N-2}, \quad (\pi M)_N = \pi_N + p\pi_{N-1}. \tag{10.3}$$

It follows that $(\pi M)_j$ depends only $[\pi_1, \ldots, \pi_{N-1}]$ for $j = 1, \ldots, N-1$. Thanks to the fact that $\pi_j \geqslant 0$ for each $j = 0, \ldots, N$ and that $\sum_{j=0}^N \pi_j = 1$, we conclude that

$$(\pi M)_j \leqslant \max(p, 1-p) = q \in (0,1) \quad \text{for } j = 1, \ldots, N-1.$$

Now, since $(\pi M)_j \geqslant 0$ for each $j = 0, \dots, N$ and since $\sum_{j=0}^{N} (\pi M)_j \leqslant 1$, the argument can be iterated to give

$$(\pi M^n)_j \leqslant q^n \quad \text{for } n \in \mathbb{N} \text{ and } j = 1, \dots, N-1.$$

Choosing $\pi = e_2, \dots, e_N$ and letting $n \to \infty$, we see that

$$(e_{i+1}^{\mathsf{T}} M^{\infty})_j = (M^{\infty})_j^i = 0 \quad \text{for } i, j = 1, \dots, N-1.$$

Since the rows of M^{∞} sum to 1, we learn that

$$w_m + l_m = (M^{\infty})_N^m + (M^{\infty})_0^m = 1,$$

i. e., that each single game ends with a winner or a loser with probability 1. Furthermore, if player A begins with m chips and wins, given a probability $p \in (0, 1)$ of flipping heads, then player B loses the same game that she begins with $N - m$ chips with her winning probability in each toss amounting to $(1 - p)$. This shows that $w_m(p) = l_{N-m}(1 - p)$ and, with the above, that $w_m(p) = 1 - w_{N-m}(1 - p)$.

10.6.3 Game duration

Another interesting question that can be asked about this game is its expected duration. A game can possibly run forever. Take $m = 2$ and $N = 2$ so that each player starts with 2 chips. If the sequence $(H, T, H, T, \dots)$ of flips is observed, then the game continues forever without a winner. We also know that the game will last at least two flips, since this is the number of consecutive losses, which would cause either player to lose. Each duration of the game has a certain probability. For the specific example, we have that

$$P(D = 2) = P(\{(H, H), (T, T)\}) = p^2 + (1 - p)^2,$$

where D denotes the duration of the game. In general, a game can have any duration k and we denote the corresponding probability by $P(D_m = k)$, where the index m indicates that we consider a game in which player A begins with m chips, while as before the total number N of chips in the game is given and fixed. We allow for the possibility that $P(D_m = k) = 0$ in order to avoid having to specify the durations of a game that are actually possible. In the specific example with $m = N = 2$, we see that

$$P(D_2 = 0) = 0, \quad P(D_2 = 1) = 0, \quad P(D_2 = 2) = p^2 + (1 - p)^2,$$
$$P(D_2 = 3) = 0, \quad P(D_2 = 4) = ?, \quad \dots$$

A histogram of observed game durations for 5,000 randomly generated games is shown below. It is assumed that player A starts with 40 of a total of 60 chips and that the coin is fair. Verify that the computed winning frequency and average game duration are close to the theoretical values we derived and are about to derive, respectively.

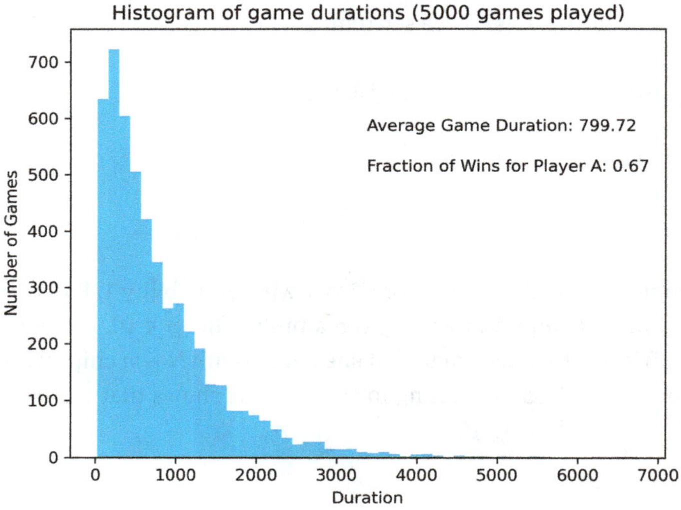

The expected duration of a game is given by

$$d_m = E[D_m] = \sum_{i=0}^{\infty} i\,P(D_m = i), \qquad m = 1, \dots, N.$$

We know that $d_0 = d_N = 0$ since the game cannot even start if player A has none or all of the chips. Next, we show that

$$d_m = 1 + p\,d_{m+1} + (1-p)\,d_{m-1},$$

for $m = 1, \dots, N-1$. This will allow us to determine d_m for each m. In order to verify the identity, observe that, for $m = 1, \dots, N-1$, it holds that

$$\begin{aligned} P(D_m = i) &= P(D_m = i|H)P(H) + P(D_m = i|T)P(T) \\ &= p\,P(D_{m+1} = i-1) + (1-p)\,P(D_{m-1} = i-1) \end{aligned}$$

by conditioning on the first flip being heads or tails by Bayes' formula. Notice that we can consider the first flip for $m = 1, \dots, N-1$ since the game will last at least one flip. We also used the fact that the duration of the game increases by 1 after the first flip and that A's chips' count increases or decreases by 1 depending on whether the first flip is heads or tails. This implies that

$$\begin{aligned}
E[D_m] &= \sum_{i=0}^{\infty} i P(D_m = i) = \sum_{i=1}^{\infty} i P(D_m = i) \\
&= \sum_{i=1}^{\infty} i[p P(D_{m+1} = i-1) + (1-p) P(D_{m-1} = i-1)] \\
&= p \sum_{i=1}^{\infty} i P(D_{m+1} = i-1) + (1-p) \sum_{i=1}^{\infty} i P(D_{m-1} = i-1) \\
&= p \sum_{j=0}^{\infty} (j+1) P(D_{m+1} = j) + (1-p) \sum_{j=0}^{\infty} (j+1) P(D_{m-1} = j) \\
&= p + (1-p) + pE[D_{m+1}] + (1-p)E[D_{m-1}],
\end{aligned}$$

since $\sum_{j=0}^{\infty} P(D_m = j) = 1$ for any m and thanks to the definition of expected value. The desired identity follows since $E[D_m] = d_m$ for any m and can be rewritten as

$$\begin{aligned}
d_{m+1} - d_m &= -\frac{1}{p} + r(d_m - d_{m-1}) = \cdots \\
&= -\frac{1}{p}(1 + r + \cdots + r^{m-1}) + r^m(d_1 - d_0) \\
&= -\frac{1}{p}\frac{1-r^m}{1-r} + r^m d_1,
\end{aligned}$$

assuming that $r \neq 1$ and where $r = \frac{1-p}{p}$. The case when $r = 1$, i. e., $p = \frac{1}{2}$, is left as an exercise. It holds that

$$\begin{aligned}
d_m &= \sum_{k=0}^{m-1} (d_{k+1} - d_k) = \sum_{k=0}^{m-1} \left[-\frac{1}{p}\frac{1-r^k}{1-r} + r^k d_1 \right] \\
&= -\frac{1}{p(1-r)} \left[m - \frac{1-r^m}{1-r} \right] + \frac{1-r^m}{1-r} d_1 \\
&= \frac{1}{1-2p} \left[\frac{1-r^m}{1-r} - m \right] + \frac{1-r^m}{1-r} d_1
\end{aligned}$$

Setting $m = N$ and using that $d_N = 0$, we arrive at

$$d_1 = \frac{N}{p(1-r^N)} - \frac{1}{2p-1},$$

and consequently, at

$$d_m = \frac{1}{1-2p} \left[m - N\frac{1-r^m}{1-r^N} \right] \quad \text{for } m = 0, \dots, N.$$

10.7 Vector-valued random variables and variance

Every variable that exist in a deterministic form will also exist in random form. In particular, we can consider *vector-valued random variables*, i. e., random variables with values in a vector space, say $\mathbb{R}^n$. These are functions of the form

$$X : S \to \mathbb{R}^n$$

defined on a probability space S. We can think of X just as well as a vector consisting of scalar random variables $X^1, \dots, X^n$, which are its components, i. e., satisfy

$$X = (X^1, \dots, X^n).$$

The corresponding expected value is the vector $E[X]$ with

$$E[X] = (E[X^1], \dots, E[X^n]).$$

The expected value $\mu = E[X]$ of a real-valued ($n = 1$) random variable is the value $\mu \in \mathbb{R}$, which makes the expected difference between X and μ vanish, i. e., such that

$$E[X - \mu] = 0 (= E[X] - \mu).$$

Unless X is deterministic (i. e., constant), it will take on multiple values. It is natural to ask how far such values are from the mean, on average. While a possible measure of mean distance to the average could be given by $E[|X - \mu|]$, for technical reasons, the preferred quantities are *variance*, which is given by

$$\operatorname{var}(X) = E[|X - \mu|^2],$$

and the related *standard deviation*

$$\operatorname{std}(X) = \sqrt{\operatorname{var}(X)} = E[|X - \mu|^2]^{\frac{1}{2}},$$

which scales like a distance. For a vector valued random variable $X \in \mathbb{R}^n$, the corresponding quantity is given by

$$\operatorname{var}(X) = E[(X - \mu)(X - \mu)^\top] \in \mathbb{R}^{n \times n},$$

and it is called *covariance matrix*. Notice how this symmetric matrix not only contains the variance

$$\operatorname{var}(X^i) = \operatorname{var}(X)_{ii}$$

of the components of the vector X on its diagonal, but also, in its off-diagonal entries, the so-called *correlations*

$$E[(X^i - \mu^i)(X^j - \mu^j)] = \text{var}(X)_{ij}$$

of the components of X. If you think of the correlation as an inner product of the corresponding random variables, then the intuition is that it would be large if both random variables simultaneously assume nonvanishing values for many of the same outcomes. In this sense, one could be used as a proxy for the other. Take for instance the binary random variables of it raining and of carrying an umbrella. While the two random variables are different since it is possible to carry an umbrella when the weather is nice or not carry one when it is raining, it is the case that these variables will both have the same value a lot of the time. Their correlation will reflect this by exhibiting a nonzero value.

10.7.1 Principal component analysis

Next, we showcase a nice connection between linear algebra (singular value decomposition), statistics/probability, and data science. It is given by the so-called Principal Component Analysis (PCA), a widely used method of data analysis.

Given a sample of data vectors $x_1, \dots, x_n \in \mathbb{R}^m$ (measurements of some kind) for which m is large, we would like to identify (a small number of) directions in $\mathbb{R}^m$ along which the data exhibit most of their variability. These could be and are interpreted as distinguishing features. The framework consists in assuming that the data are a manifestation of the values of a random vector $X \in \mathbb{R}^m$ in the sense that the frequency of values in the observed data set reflects the probability of corresponding values of X. With this understanding, the value

$$\overline{x} = \frac{1}{n} \sum_{i=1}^{n} x_i$$

can be thought of as a reasonable guess for the expected value $E[X]$ based on the observed data. It is called *sample mean* for apparent reasons. As the data points x_i can be replaced by $x_i - \overline{x}$ with no loss of information, we may assume, without loss of generality, that $\overline{x} = 0$. Let us collect the data in the so-called design matrix $\mathbb{X} \in \mathbb{R}^{n \times m}$, which features the data vectors in its n rows $\mathbb{X}^i_\bullet$, $i = 1, \dots, n$. Then it is natural to interpret $\mathbb{X}^\top \mathbb{X}$ as an approximation (up to a multiplicative constant) of the covariance matrix $E[XX^\top]$ of X since it holds that

$$\begin{aligned}(\mathbb{X}^\top \mathbb{X})^j_k &= \sum_{i=1}^{n} (\mathbb{X}^\top)^j_i \mathbb{X}^i_k = \sum_{i=1}^{n} \mathbb{X}^i_j \mathbb{X}^i_k = \sum_{i=1}^{n} x^j_i x^k_i \\ &= \Big(\sum_{i=1}^{n} x_i x_i^\top \Big)^j_k = \Big(\sum_{i=1}^{n} (x_i - 0)(x_i - 0)^\top \Big)^j_k\end{aligned}$$

Here, we assume that X has also been centered (i. e., been replaced by $X - \mu$) so that $E[X] = 0$ without loss of generality. We know that the optimal approximation of $\mathbb{X}$ (in the sense of the Frobenius norm as explained in Section 8.1) by a rank-k matrix is given by its truncated singular value decomposition that keeps only the k largest singular values. Let $K = \operatorname{rank}(A) \leqslant \min(n, m)$ and

$$\mathbb{X} = \sum_{j=1}^{K} \sigma_j v_j u_j^\top = V\Sigma U^\top$$

be the singular value decomposition of $\mathbb{X}$. Recall that

$$\sigma_1 \geqslant \sigma_2 \geqslant \cdots \geqslant \sigma_K > 0$$

and that the vectors $v_1, \dots, v_K \in \mathbb{R}^n$ can be constructed to build an orthonormal system and so can the vectors $u_1, \dots, u_K \in \mathbb{R}^m$. We compute

$$\mathbb{X}^\top \mathbb{X} = U\Sigma V^\top V\Sigma U^\top = U\Sigma^2 U^\top,$$

which can be interpreted as saying that the covariance matrix of the transformed (projected) data

$$(U^\top \mathbb{X}^\top)_1^\bullet = U^\top x_1, \dots, (U^\top \mathbb{X}^\top)_n^\bullet = U^\top x_n \in \mathbb{R}^K$$

with design matrix $\mathbb{X}U$ is diagonal as follows from $\mathbb{X}U = V\Sigma$ and

$$(\mathbb{X}U)^\top \mathbb{X}U = \Sigma^\top V^\top V\Sigma = \Sigma^2.$$

Letting $U_k \in \mathbb{R}^{m\times k}$ be the submatrix consisting of the first k columns of U (corresponding to the k-largest variances $\sigma_1^2, \dots, \sigma_k^2$), we obtain the directions $u_j = U_j^\bullet$ along which the data have variance σ_j^2 for $j = 1, \dots, k$. Verify (Q4) that this means that the scalar quantities $u_j^\top x_i$, for $i = 1, \dots, n$, exhibit variance amounting to σ_j^2. Notice also that the projected data $U_k^\top x_1, \dots, U_k^\top x_n$ are uncorrelated as the off-diagonal entries of the corresponding covariance matrix vanish, a fact that follows from the identity $\mathbb{X}U_k = V_k\Sigma_k$, the validity of which rests on

$$\mathbb{X}U_k U_k^\top = \mathbb{X}\sum_{j=1}^{k} u_j u_j^\top = \sum_{j=1}^{k} \sigma_j v_j u_j^\top = V_k \Sigma_k U_k^\top,$$

which holds for any $k \leqslant K$ (the case $k = K$ was studied above) and where V_k is the matrix consisting of the first k columns of V and $\Sigma_k \in \mathbb{R}^{k\times k}$ is the diagonal matrix with the k largest variances on its diagonal.

In the two-dimensional example depicted below, a sample of data points $x_1, \ldots, x_{50} \in \mathbb{R}^2$ is plotted along with the direction (blue dotted line) in which the most variability is observed. This direction is determined by the vector u_1 that corresponds to the largest singular value σ_1 of the design matrix $\mathbb{X}$. In this direction, the data exhibits a sample variance σ_1^2.

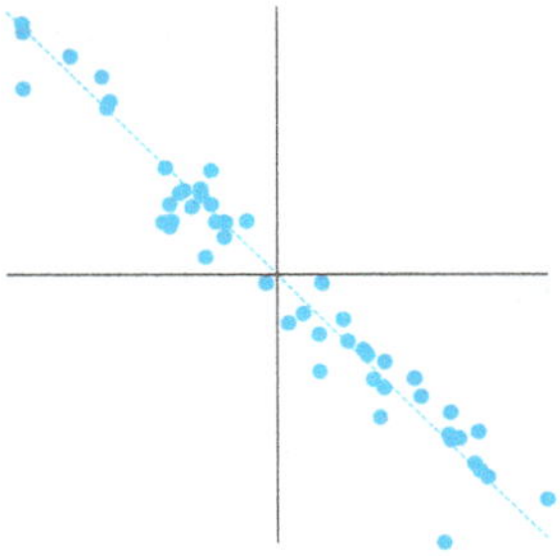

10.8 Concluding remarks

In this chapter, we learned that conditional probability is a powerful tool and that the computation of many a probability is intimately connected to (partial) differential and difference equations. We chose to present prototypical examples in a discrete context in order to avoid technicalities, which may have obfuscated the core ideas. It is, however, possible and interesting to study phenomena with continuous time and/or continuous sample space.

11 Answers to selected questions

11.1 Answer to Chapter 1 question

Answer to Q1

If G is the graph of a map $m : S \to T$ with inverse m^{-1}, then the graph of the inverse is

$$G(m^{-1}) = \{(t,s) \mid (s,t) \in G\} \subset T \times S.$$

In other words, the graph of the inverse is obtained by flipping the order in the pairs of the graph of m.

11.2 Answers to Chapter 2 questions

Answer to Q1

Let $s \in S$. Due to **(er1)**, we have that $s \in [s]$. This shows that

$$S \subset \bigcup_{s \in S} [s].$$

Next, take $[s] \neq [t] \in S/\sim$ and assume that there is $r \in S$ such that $r \in [s] \cap [t]$, i. e., that $r \sim s$ and $r \sim t$. Then, by **(er2)** $s \sim r$, so that $s \sim t$ by **(er3)** and we would have to conclude $s \in [t]$ and $t \in [s]$. This is, however, not the case since $[s] \neq [t]$ and we conclude that $[s] \cap [t] = \emptyset$.

Answer to Q2

Let $(m,n) \in \mathbb{N}^2$ and observe that $m \cdot n = n \cdot m$ showing that **(er1)** is satisfied. Next, $(m,n) \sim (p,q)$ means $m \cdot q = n \cdot p$, which, using commutativity, rewrites as $p \cdot n = q \cdot m$ and means $(p,q) \sim (m,n)$ and yields $(p,q) \sim (m,n)$. This gives **(er2)**. Finally, let $(m,n) \sim (p,q)$ and $(p,q) \sim (r,s)$, which amount to

$$m \cdot q = n \cdot p \quad \text{and} \quad p \cdot s = q \cdot r.$$

It follows that

$$(m \cdot s) \cdot (p \cdot q) = (m \cdot q) \cdot (p \cdot s) = (n \cdot p) \cdot (q \cdot r) = (n \cdot r) \cdot (p \cdot q),$$

and simplifying that $m \cdot s = n \cdot r$. This means $(m,n) \sim (r,s)$ and gives **(er3)**, thus concluding the proof.

https://doi.org/10.1515/9783110780925-011

Answer to Q3

It is simply $[n, m]$ since

$$[n, m] \cdot [m, n] = [n \cdot m, m \cdot n] = [n \cdot m, n \cdot m] = [1, 1] = 1,$$
$$[m, n] \cdot [n, m] = [m \cdot n, n \cdot m] = [m \cdot n, m \cdot n] = [1, 1] = 1.$$

Answer to Q4

Notice that $(\mathbb{N}, +)$, just as $(\mathbb{N}, \cdot)$, enjoys commutativity and associativity. These were the only two properties used in order to show that $(m, n) \sim (p, q)$ defined by $m \cdot q = n \cdot p$ is an equivalence relation on $\mathbb{N}^2$. The claim therefore follows by replacing $\cdot$ by $+$.

Answer to Q5

Well-defined: The proof is again obtained by replacing $\cdot$ in the corresponding argument that was used in the construction of $\mathbb{Q}^+$ by $+$ and observing that only commutativity and associativity are needed.

Extension: The natural numbers m and n correspond to the new numbers $[n+1, 1]$ and $[m + 1, 1]$ so that, according to the new definition of addition, their sum is

$$[n + 1, 1] + [m + 1, 1] = [n + m + 2, 2] = [n + m + 1, 1],$$

which corresponds to $n + m$ as desired. *Commutativity:* It holds that

$$[m, n] + [k, l] = [m + k, n + l] = [k + m, l + n] = [k, l] + [m, n]$$

since we can use commutativity in each component, that is commutativity of $(\mathbb{N}, +)$.

Associativity: Similarly, we can use the associativity of $(\mathbb{N}, +)$ in each component to see that

$$\begin{aligned}([m, n] + [k, l]) + [i, j] &= [m + k, n + l] + [i, j] = \left[(m + k) + i, (n + l) + j\right] \\ &= \left[m + (k + i), n + (l + j)\right] \\ &= [m, n] + ([k, l] + [i, j])\end{aligned}$$

Answer to Q6

Multiplication on $\mathbb{Z}$ can be defined by

$$[m, n] \cdot [k, l] = [m \cdot k + n \cdot l, m \cdot l + n \cdot k],$$

which amounts to the familiar

$$(m - n) \cdot (k - l) = (m \cdot k + n \cdot l) - (m \cdot l + n \cdot k).$$

Notice how we have that

$$\begin{aligned}(-1) \cdot n = [1, 2] \cdot [n + 1, 1] &= [n + 1 + 2, 1 + 2 \cdot (n + 1)] \\ &= [1, n + 1] = -n.\end{aligned}$$

Answer to Q7

Starting with $\mathbb{Z}$, we can introduce the equivalence relation given by

$$(m, n) \sim (k, l) \quad \text{iff} \quad m \cdot l = n \cdot k,$$

on $\mathbb{Z} \times \mathbb{N}$ in order to obtain $\mathbb{Q} = (\mathbb{Z} \times \mathbb{N})/{\sim}$. Starting with $\mathbb{Q}^+$, we can alternatively define the equivalence relation

$$(p, q) \sim (r, s) \quad \text{iff} \quad p + s = q + r,$$

and obtain $\mathbb{Q} = (\mathbb{Q}^+ \times \mathbb{Q}^+)/{\sim}$.

Answer to Q8

We only verify the inequality using the fact that $x \leqslant |x|$ for any $x \in \mathbb{Q}$, which holds since $x = |x|$ if $x \in \mathbb{Q}^+$ and $x \leqslant -x$ if $x \in \mathbb{Q}^-$. If $x - z \in \mathbb{Q}^+$, then

$$|x - z| = x - z = x - y + y - z \leqslant |x - y| + |y - z|,$$

if $x - z \in \mathbb{Q}^-$, then

$$|x - z| = z - x = z - y + y - x \leqslant |z - y| + |y - x| = |x - y| + |y - z|,$$

and if $x - z = 0$, then

$$|x - z| \leqslant |x - y| + |y - z|.$$

We also used that $|x| = |-x|$ as can be easily verified.

Answer to Q9

Let the sequence $(x_n)_{n\in\mathbb{N}}$ of rationals converge to $x_\infty \in \mathbb{Q}$ and let $M \in \mathbb{N}$ be given. The definition of convergence yields $N \in \mathbb{N}$ such that

$$|x_n - x_\infty| \leqslant \frac{1}{2M} \quad \text{for } n \geqslant N,$$

from which we conclude that

$$|x_n - x_m| \leqslant |x_n - x_\infty| + |x_\infty - x_m| \leqslant \frac{1}{2M} + \frac{1}{2M} = \frac{1}{M} \quad \text{for } m, n \geqslant N.$$

Thus the sequence is Cauchy.

Answer to Q10

If $(x_n)_{n\in\mathbb{N}}$ has the positivity property, i. e., if there exist $M \in \mathbb{N}$ and $N \in \mathbb{N}$ such that

$$x_n \geqslant \frac{1}{M} \quad \text{for } n \geqslant N,$$

and if $(y_n)_{n\in\mathbb{N}}$ is an equivalent Cauchy sequence, we can find $\widetilde{N} \in \mathbb{N}$ such that

$$|y_n - x_n| \leqslant \frac{1}{2M} \quad \text{for } n \geqslant \widetilde{N}.$$

It follows that

$$y_n = y_n - x_n + x_n \geqslant x_n - |y_n - x_n| \geqslant \frac{1}{2M},$$

and thus that the stated positivity condition holds for all Cauchy sequences in the equivalence class of $(x_n)_{n\in\mathbb{N}}$ with lower bound $\frac{1}{2M}$.

Answer to Q11

It is enough to notice that

$$|q_n - q_m| \leqslant |q_n - x_n| + |x_n - x_m| + |x_m - q_m| \leqslant |x_n - x_m| + \frac{1}{n} + \frac{1}{m},$$

and to use the assumption that $(x_n)_{n\in\mathbb{N}}$ is a Cauchy sequence. Indeed, given $\varepsilon \in \mathbb{Q}^+$, we can find $N \in \mathbb{N}$ such that

$$|x_n - x_m| \leqslant \frac{\varepsilon}{3}, \quad \frac{1}{k} \leqslant \frac{\varepsilon}{3} \quad \text{for } k, m, n \geqslant N,$$

from which we conclude that

$$|q_n - q_m| \leqslant \frac{\varepsilon}{3} + \frac{\varepsilon}{3} + \frac{\varepsilon}{3} = \varepsilon \quad \text{for } m, n \geqslant N,$$

using the first inequality.

11.3 Answers to Chapter 3 questions

Answer to Q1

First, observe that $\mathbb{C}^{\mathbb{N}}_{cs} = \mathbb{C}^{\mathbb{N}}_{c}$ since $\mathbb{C}$ is complete so that Cauchy sequences are convergent, while convergent sequences are always Cauchy. If two sequences $x = (x_n)_{n\in\mathbb{N}}$ and $y = (y_n)_{n\in\mathbb{N}}$ converge (with limits x_∞ and y_∞), then, given $\varepsilon > 0$ and $\lambda \in \mathbb{C}$, we can find $N \in \mathbb{N}$ with

$$|x_n - x_\infty| \leqslant \frac{\varepsilon}{2\max\{|\lambda|, 1\}} \quad \text{and} \quad |y_n - y_\infty| \leqslant \frac{\varepsilon}{2} \quad \text{for } n \geqslant N.$$

Such N exists for each individual sequence by the definition of convergence and taking the maximum yields one that works for both. It follows that

$$\big|x_n + y_n - (x_\infty + y_\infty)\big| \leqslant |x_n - x_\infty| + |y_n - y_\infty| \leqslant \varepsilon \quad \text{for } n \geqslant N,$$

which yields the convergence of $(x_n + y_n)_{n\in\mathbb{N}}$ to $x_\infty + y_\infty$ and shows that $x + y \in \mathbb{C}^{\mathbb{N}}_{c}$. It also follows that

$$|\lambda x_n - \lambda x_\infty| \leqslant |\lambda||x_n - x_\infty| \leqslant \varepsilon \quad \text{for } n \geqslant N,$$

which gives the convergence of $(\lambda x_n)_{n\in\mathbb{N}}$ to the limit λx_∞ and shows that $\lambda x \in \mathbb{C}^{\mathbb{N}}_{c}$, concluding the proof that $\mathbb{C}^{\mathbb{N}}_{c}$ is a subspace of $\mathbb{C}^{\mathbb{N}}$.

Answer to Q2

Assume that $V = \operatorname{span}\{u_1, \dots, u_n\}$ and $V = \operatorname{span}\{v_1, \dots, v_m\}$. Then there are $a_j^1, \dots, a_j^n \in \mathbb{F}$ for $j = 1, \dots, m$ as well as $b_k^1, \dots, b_k^m \in \mathbb{F}$ for $k = 1, \dots, n$ such that

$$a_j^1 u_1 + \cdots + a_j^n u_n = v_j \quad \text{and} \quad b_k^1 v_1 + \cdots + b_k^m v_m = u_k$$

for $j = 1, \dots, m$ and $k = 1, \dots, n$. For any vector $u \in V$, we can find $\alpha \in \mathbb{F}^n$ and $\beta \in \mathbb{F}^m$ such that

$$u = \sum_{j=1}^{n} \alpha^j u_j = \sum_{j=1}^{n} \alpha^j \sum_{k=1}^{m} b_j^k v_k = \sum_{k=1}^{m} \left[\sum_{j=1}^{n} b_j^k \alpha^j \right] v_k$$

$$= \sum_{k=1}^{m} \beta^k v_k = \sum_{k=1}^{m} \beta^k \sum_{j=1}^{n} a_k^j u_j = \sum_{j=1}^{n} \left[\sum_{k=1}^{m} a_k^j \beta^k \right] u_j.$$

It follows from the definition of basis that

$$\alpha^j = \sum_{k=1}^{m} a_k^j \beta^k \quad \text{for } j = 1, \ldots, n \quad \text{and} \quad \beta^k = \sum_{j=1}^{n} b_j^k \alpha^j \quad \text{for } k = 1, \ldots, m.$$

This shows that the matrices $A = [a_k^j]_{k=1,\ldots,m}^{j=1,\ldots,n}$ and $B = [b_j^k]_{j=1,\ldots,n}^{k=1,\ldots,m}$ are inverse to each other. We, however, know (from basic linear algebra) that matrices can only be invertible if they are square and this requires $m = n$.

Answer to Q3

We give a proof for the functions f_y. Let $y_1, \ldots, y_n$ be distinct points in $\mathbb{R}$ for arbitrary $n \in \mathbb{N}$ and assume that

$$f = \sum_{j=1}^{n} \alpha^j f_{y_j} \equiv 0.$$

Then $f(x) = 0$ for every $x \in \mathbb{R}$ and, in particular,

$$0 = f(y_k) = \sum_{j=1}^{n} \alpha^j f_{y_j}(y_k) = \alpha^k \quad \text{for } k = 1, \ldots, n.$$

This shows that the vectors (functions) $f_{y_1}, \ldots, f_{y_n}$ are linearly independent as desired.

Answer to Q4

Starting with the function lim, notice that linearity amounts to

$$\lim_{n\to\infty} (\alpha x_n + \beta y_n) = \alpha x_\infty + \beta y_\infty,$$

for sequences $(x_n)_{n\in\mathbb{N}}, (y_n)_{n\in\mathbb{N}}$ converging to x_∞ and y_∞, respectively. This is precisely what we proved in a previous answer. As for δ_{x_0}, the claim follows from

$$\delta_{x_0}(\alpha f + \beta g) = (\alpha f + \beta g)(x_0) = \alpha f(x_0) + \beta g(x_0) = \alpha \delta_{x_0}(f) + \beta \delta_{x_0}(g),$$

which is valid for any functions $f, g \in \mathbb{R}^{\mathbb{R}}$, and scalars $\alpha, \beta \in \mathbb{R}$.

Answer to Q5

We need to show that

$$\mathcal{L}(V, W) = \{\mathcal{L} : V \to W \mid \mathcal{L} \text{ is linear}\}$$

is a vector space. Notice first that any set of functions W^S from a set S into a vector space is itself a vector space with respect with pointwise addition and pointwise scalar multiplication given by

$$(f+g)(s) = f(s) + g(s) \quad \text{and} \quad (\lambda f)(s) = \lambda f(s) \quad \text{for } s \in S,$$

for $f, g \in W^S$, and $\lambda \in \mathcal{F}$. Then the claim follows observing that $\mathcal{L}(V, W)$ is a subspace of W^V since the sum of any two linear maps is linear as is any scalar multiple of a linear map.

Answer to Q6

We shall use the so-called Hölder's inequality, which states that

$$\sum_{j=1}^{m} |\alpha|^j |\beta|^j \leqslant |\alpha|_p |\beta|_{p'} \quad \text{for } \alpha, \beta \in \mathbb{F}^m,$$

where $p \in [1, \infty]$ and $\frac{1}{p} + \frac{1}{p'} = 1$. To derive this inequality, we use that

$$xy \leqslant \frac{1}{p} x^p + \frac{1}{p'} y^{p'} \quad \text{for } x, y \geqslant 0,$$

the validity of which is obtained exploiting the concavity of the logarithm. Indeed, we have that

$$\log\left(\frac{1}{p} x^p + \frac{1}{p'} y^{p'}\right) \geqslant \frac{1}{p} \log(x^p) + \frac{1}{p'} \log(y^{p'}) = \log(xy),$$

which gives the inequality after exponentiation. To obtain Hölder's inequality, it is then enough to replace α by $\widetilde{\alpha} = \frac{\alpha}{|\alpha|_p}$ and β by $\widetilde{\beta} = \frac{\beta}{|\beta|_q}$, which is possible if $\alpha \neq 0 \neq \beta$. Notice that Hölder's inequality holds true when $\alpha = 0$ or $\beta = 0$ and we can therefore focus on the case α and β do not vanish. Then

$$|\widetilde{\alpha}^j \widetilde{\beta}^j| \leqslant \frac{1}{p} |\widetilde{\alpha}^j|^p + \frac{1}{p'} |\widetilde{\beta}^j|^{p'} \quad \text{for } j = 1, \ldots, m,$$

and adding

$$\frac{1}{|\alpha|_p|\beta|_q}\sum_{j=1}^{m}|\alpha^j\beta^j| = \sum_{j=1}^{m}|\tilde{\alpha}^j\tilde{\beta}^j| \leqslant \frac{1}{p}\sum_{j=1}^{m}|\tilde{\alpha}^j|^p + \frac{1}{p'}\sum_{j=1}^{m}|\tilde{\beta}^j|^{p'}$$
$$= \frac{1}{p} + \frac{1}{p'} = 1,$$

since $\sum_{j=1}^m |\tilde{\alpha}^j|^p = 1 = \sum_{j=1}^m |\tilde{\beta}^j|^{p'}$ by construction. Now take $p \in (0, \infty)$ and observe that

$$|\alpha|_\infty^p = \Big(\max_{i=1,\dots,m}|\alpha^i|\Big)^p = \max_{i=1,\dots,m}|\alpha^i|^p \leqslant \sum_{j=1}^{m}|\alpha^j|^p = |\alpha|_p^p$$

and that

$$|\alpha|_p^p = \sum_{j=1}^{m}|\alpha^j|^p \leqslant m \max_{i=1,\dots,m}|\alpha^i|^p = m|\alpha|_\infty^p,$$

for $\alpha \in \mathbb{F}^m$. This amounts to

$$|\alpha|_\infty \leqslant |\alpha|_p \leqslant m^{1/p}|\alpha|_\infty, \quad \alpha \in \mathbb{F}^m.$$

Next, take $1 \leqslant p < q < \infty$ and observe that

$$|\alpha|_q^q = \sum_{j=1}^{m}|\alpha^j|^q \leqslant \sum_{j=1}^{m}|\alpha|_\infty^{q-p}|\alpha^j|^p \leqslant |\alpha|_\infty^{q-p}|\alpha|_p^p \leqslant |\alpha|_p^q$$

by the first inequality for $|\cdot|_\infty$ above, and that

$$|\alpha|_p^p = \sum_{j=1}^{m} 1\cdot|\alpha|^p \leqslant \Big(\sum_{j=1}^{m}1^r\Big)^{1/r}\Big(\sum_{j=1}^{m}|\alpha^j|^{pr'}\Big)^{1/r'} = m^{p/q}|\alpha|_q^p,$$

by Hölder's inequality with $r' = \frac{q}{p}$. The last two inequalities amount to

$$|\alpha|_q \leqslant |\alpha|_p \leqslant m^{1/p-1/q}|\alpha|_q, \quad \alpha \in \mathbb{F}^m.$$

Answer to Q7

Let $p \in [1, \infty)$ and $(x_n)_{n\in\mathbb{N}}$ be a Cauchy sequence in $\mathbb{R}_p^{\mathbb{N}}$. Fixing $k \in \mathbb{N}$ and considering a single component $(x_n^k)_{n\in\mathbb{N}}$ we obtain a Cauchy sequence of real numbers since

$$|x_n^k - x_m^k| \leqslant \Big(\sum_{j\in\mathbb{N}}|x_n^j - x_m^j|^p\Big)^{1/p} = |x_n - x_m|_p.$$

Completeness of $\mathbb{R}$ yields a limit x_∞^k for each $k \in \mathbb{N}$, i. e. a sequence $x_\infty = (x_\infty^k)_{k\in\mathbb{N}}$ in $\mathbb{R}^\mathbb{N}$, which may or may not be an element of $\mathbb{R}_p^\mathbb{N}$. We prove next that it is. Notice that, given $\varepsilon > 0$, there is $N \in \mathbb{N}$ such that

$$\left(\sum_{k=1}^{M} |x_n^k - x_m^k|^p\right)^{1/p} \leqslant \left(\sum_{k=0}^{\infty} |x_n^k - x_m^k|^p\right)^{1/p} = |x_n - x_m|_p \leqslant \varepsilon \quad \text{for } n, m \geqslant N,$$

and we can let m tend to ∞ in the first sum exploiting the convergence of each component individually to obtain

$$\left(\sum_{k=1}^{M} |x_n^k - x_\infty^k|^p\right)^{1/p} \leqslant \varepsilon \quad \text{for } n \geqslant N.$$

The latter inequality is valid for any finite M and, letting $M \to \infty$, it is inferred that $|x_n - x_\infty|_p \leqslant \varepsilon$ for $n \geqslant N$. This shows that $x_\infty \in \mathbb{R}_p^\mathbb{N}$ since

$$|x_\infty|_p \leqslant |x_N|_p + \varepsilon,$$

and that $|x_n - x_\infty|_p \to 0$ as $n \to \infty$ (since $\varepsilon > 0$ is arbitrary), thus concluding the proof. The last inequality follows from the validity of

$$\big||y|_p - |z|_p\big| \leqslant |y - z|_p \quad \text{for } y, z \in \mathbb{R}_p^\mathbb{N},$$

which you should verify.[1]

Answer to Q8

This metric generates the topology of componentwise convergence. To see this, we take a sequence $(x_n)_{n\in\mathbb{N}}$ of sequences in $\mathbb{R}^\mathbb{N}$ where each term $x_n \in \mathbb{R}^\mathbb{N}$ in this sequence has components x_n^j for $j \in \mathbb{N}$ and assume that

$$\lim_{n\to\infty} x_n^j = x_\infty^j \quad \text{for every } j \in \mathbb{N}.$$

This means that, given any $\varepsilon > 0$, we can find a number $N = N(j) \in \mathbb{N}$ such that

$$|x_n^j - x_\infty^j| \leqslant \varepsilon/2 \quad \text{provided } n \geqslant N(j).$$

Simultaneously, we can find $M \in \mathbb{N}$ such that

1 Use the triangle inequality.

$$\sum_{j=M+1}^{\infty} \frac{|x_n^j - x_\infty^j|}{1 + |x_n^j - x_\infty^j|} 2^{-j} \leqslant \sum_{j=M+1}^{\infty} 2^{-j} \leqslant \varepsilon/2,$$

since the series $\sum_{j\in\mathbb{N}} 2^{-j}$ converges. Taking $N = \max\{N(1), \dots, N(M)\}$, it follows that

$$\begin{aligned} d(x_n, x_\infty) &= \sum_{j\in\mathbb{N}} \frac{|x_n^j - x_\infty^j|}{1 + |x_n^j - x_\infty^j|} 2^{-j} \leqslant \sum_{1\leqslant j\leqslant M} \frac{|x_n^j - x_\infty^j|}{1 + |x_n^j - x_\infty^j|} 2^{-j} + \varepsilon/2 \\ &\leqslant \sum_{1\leqslant j\leqslant M} \varepsilon 2^{-j-1} + \varepsilon/2 \leqslant \varepsilon \quad \text{if } n \geqslant N. \end{aligned}$$

This shows that componentwise convergence implies convergence with respect to d. If, on the other hand, we assume convergence in d, then for any given $\delta > 0$, we can find $N \in \mathbb{N}$ such that

$$d(x_n, x_\infty) = \sum_{j\in\mathbb{N}} \frac{|x_n^j - x_\infty^j|}{1 + |x_n^j - x_\infty^j|} 2^{-j} \leqslant \delta \quad \text{for } n \geqslant N.$$

Thus, for any given fixed $j \in \mathbb{N}$, it holds that

$$\frac{|x_n^j - x_\infty^j|}{1 + |x_n^j - x_\infty^j|} \leqslant 2^j \delta \quad \text{for } n \geqslant N.$$

Given $j \in \mathbb{N}$ and $\varepsilon > 0$, $\delta > 0$ can be chosen such that $2^j\delta < \varepsilon$ and we obtain that

$$\lim_{n\to\infty} \frac{|x_n^j - x_\infty^j|}{1 + |x_n^j - x_\infty^j|} = 0.$$

This is, however, equivalent to $\lim_{n\to\infty} |x_n^j - x_\infty^j| = 0$ since the function g defined by $g(s) = \frac{s}{1+s}$ for $s \geqslant 0$ is monotone increasing as follows by computing its derivative.

Answer to Q9

Notice that, for any $x \in \mathbb{R}^{\mathbb{N}}$ and $t > 0$, it holds that

$$d(tx, 0) = \sum_{n\in\mathbb{N}} \frac{|tx_n|}{1 + |tx_n|} 2^{-n} \leqslant \sum_{n\in\mathbb{N}} 2^{-n} = 1.$$

If d were induced by a norm $|\cdot|$, it would have to hold that

$$t|x| = |tx| = |tx - 0| = d(tx, 0) \leqslant 1 \quad \text{for every } x \in \mathbb{R}^{\mathbb{N}} \text{ and } t > 0,$$

which is clearly impossible since it would imply that $x = 0$ for every x as is seen from $|x| \leqslant t^{-1}$ by letting $t \to \infty$.

Answer to Q10

It clearly holds that $\|f\|_p \geqslant 0$ for every $f \in C([a,b],\mathbb{F})$ since the integral of nonnegative functions is nonnegative. Assume now that $\|f\|_p = 0$ but that $|f| \not\equiv 0$. The latter entails the existence of $x \in [a,b]$ with $|f(x)| > 0$. Since f and the absolute value/modulus are continuous, we can find $\delta > 0$ such that

$$|f(y)| \geqslant |f(x)|/2 \quad \text{for } y \in [x-\delta, x+\delta] \cap [a,b].$$

This implies that

$$\int_a^b |f(\xi)|^p\, d\xi \geqslant \int_{\max(a,x-\delta)}^{\min(b,x+\delta)} |f(\xi)|^p\, d\xi \geqslant \frac{\delta}{2^p}|f(x)|^p > 0.$$

It follows that $\|f\|_p = 0$ can only hold for $f \equiv 0$. Let now $\lambda \in \mathbb{F}$. The validity of $|\lambda f(\xi)| = |\lambda||f(\xi)|$ for $\xi \in [a,b]$ and the linearity of the Riemann integral implies that

$$\|\lambda f\|_p^p = \int_a^b |\lambda f(\xi)|^p\, d\xi = |\lambda|^p \int_a^b |f(\xi)|^p\, d\xi = |\lambda|^p \|f\|_p^p,$$

which is the scaling property if a norm. Finally, we consider the triangle inequality for the proof of which we need Hölder's inequality in integral form

$$\|fg\|_1 = \int_a^b |f(x)g(x)|\, dx \leqslant \|f\|_p \|g\|_{p'},$$

which is valid for $f,g \in C([a,b],\mathbb{F})$, $p \in [1,\infty)$, and $p' = \frac{p}{p-1}$ ($p' = \infty$ if $p = 1$). We first prove the triangle inequality and then Hölder's inequality. It is straightforward to verify the triangle inequality for $p = 1$ or $p = \infty$. Let therefore $p \in (1,\infty)$. It holds that

$$\begin{aligned}\|f+g\|_p^p &= \int_a^b |f(\xi)+g(\xi)||f(\xi)+g(\xi)|^{p-1}\, d\xi \\ &\leqslant (\|f\|_p + \|g\|_p)\Big(\int_a^b |f(\xi)+g(\xi)|^p\, d\xi\Big)^{\frac{p-1}{p}} = (\|f\|_p + \|g\|_p)\|f+g\|_p^{p-1},\end{aligned}$$

from which the triangle inequality follows. In this context, it is also known as Minkowski inequality. Returning to Hölder's inequality and observing that it holds whenever f or g vanish, we assume that $\|f\|_p \neq 0 \neq \|g\|_{p'}$. Using the scalar inequality

$xy \leqslant \frac{1}{p}x^p + \frac{1}{p'}x^{p'}$ valid for nonnegative numbers, we see that

$$\frac{1}{\|f\|_p\|g\|_{p'}}\int_a^b |f(\xi)g(\xi)|\,d\xi \leqslant \int_a^b \left[\frac{1}{p}\frac{|f(\xi)|^p}{\|f\|_p^p} + \frac{1}{p'}\frac{|g(\xi)|^{p'}}{\|g\|_{p'}^{p'}}\right] d\xi = \frac{1}{p} + \frac{1}{p'} = 1,$$

which concludes the proof.

Answer to Q11

It needs to be verified that $\lambda = 0$ is the only solution of

$$\sum_{j=1}^{n} \lambda^j u_j = 0.$$

Using orthogonality, we see that

$$\left(u_k \middle| \sum_{j=1}^{n} \lambda^j u_j\right) = \lambda_k (u_k | u_k) \quad \text{for every } k = 1, \dots, n.$$

It follows that $\lambda_k = 0$ for every k since the vectors are nontrivial.

11.4 Answers to Chapter 4 questions

Answer to Q1

If $z \in \mathbb{B}(x, r)$ for $r > 0$, then $|x - z| < r$ and

$$\mathbb{B}(z, r - |z - x|) \subset \mathbb{B}(x, r).$$

Indeed, for any $y \in \mathbb{B}(z, r - |z - x|)$, it holds that

$$|x - y| \leqslant |x - z| + |z - y| < |x - z| + r - |x - z| = r.$$

Answer to Q2

The sequence $(x_n)_{n\in\mathbb{N}}$ in $[0, 1)$ given by $x_n = 1 - 1/n$ has limit

$$x_\infty = 1 \notin [0, 1).$$

Answer to Q3

The set $f(C) \subset \mathbb{R}$ is compact since C is and f is continuous. This means that the extremal values $M = \sup f(C)$ and $m = \inf f(C)$ exist, are finite since compact sets are bounded, and $m, M \in f(C)$ since compact sets are closed. It follows that $x_m, x_M \in C$ can be found such that

$$\max_{x \in C} f(x) = \max f(C) = \sup f(C) = f(x_M) \quad \text{and}$$
$$\min_{x \in C} f(x) = \min f(C) = \inf f(C) = f(x_m).$$

Answer to Q4

Let $(\tilde{x}_n)_{n\in\mathbb{N}}$ be another sequence in D that converges to x. It follows from uniform continuity that

$$d_N(f(\tilde{x}_n), f(x_n)) \leqslant \varepsilon \quad \text{for } n \geqslant L,$$

since $d_M(\tilde{x}_n, x_n) \leqslant \delta$ for $n \geqslant L$ and some $L \in \mathbb{N}$. This shows that

$$\lim_{n\to\infty} d_N(f(\tilde{x}_n), f(x_n)) = 0,$$

and thus that $\lim_{n\to\infty} f(\tilde{x}_n) = \lim_{n\to\infty} f(x_n)$ as desired.

Answer to Q5

Let g be another uniformly continuous extension of f. It then holds that $\bar{f}|_D = f = g|_D$, so take $x \in M \setminus D$ and a sequence $(x_n)_{n\in\mathbb{N}}$ in D, which converges to x. It follows that

$$\bar{f}(x_n) = f(x_n) = g(x_n) \quad \text{for every } n \in \mathbb{N},$$

and that

$$g(x) = \lim_{n\to\infty} g(x_n) = \lim_{n\to\infty} f(x_n) = \lim_{n\to\infty} \bar{f}(x_n) = \bar{f}(x).$$

The claim is now a consequence of the fact that x is arbitrary.

Answer to Q6

The example shows that a bounded sequence $(f_n)_{n\in\mathbb{N}}$ of continuous functions, for which it always holds that $\{f_n \mid n \in \mathbb{N}\}$ is closed with respect to the supremum norm,[2] does not need to have a convergent subsequence. The reason is that continuity can deteriorate with increasing index, and even if the values remain in a bounded set, continuity can be lost. In the example, this happens because a jump appears in the limit. The fix consists in making sure that all terms in the sequence share a common degree of continuity. This leads to the concept of *equicontinuity*, which for a family $\mathcal{F} \subset C([-1,1])$ of continuous functions means

$$\begin{aligned}&\forall x_0 \in [-1,1],\ \forall \varepsilon > 0\ \exists \delta > 0 \quad \text{s.t.}\\ &|f(x) - f(x_0)| \leqslant \varepsilon \quad \text{when } |x - x_0| \leqslant \delta \text{ and for every } f \in \mathcal{F}.\end{aligned}$$

Equicontinuity therefore holds when "the δ needed for a given ε can be chosen to work for all $f \in \mathcal{F}$ simultaneously". *Uniform equicontinuity* holds when the above condition is satisfied with the same δ for all arguments $x_0 \in [-1,1]$. You should try to show that a bounded and uniformly equicontinuous sequence of continuous functions defined on $[-1,1]$ does indeed have a convergent subsequence. A general version of this result is known as the *Arzéla–Ascoli theorem.*

Answer to Q7

We need to show that, given $x \in \mathbb{R}_0^{\mathbb{N}}$, we can approximate it with arbitrary precision with elements of $\mathbb{R}_{00}^{\mathbb{N}}$. Since x is a null sequence, given $\varepsilon > 0$, we can find $N \in \mathbb{N}$ such that

$$|x_n| \leqslant \varepsilon \quad \text{for } n \geqslant N.$$

This, however, means that

$$|x^m - x|_\infty = \sup_{n\in\mathbb{N}} |x_n^m - x_n| \leqslant \varepsilon \quad \text{for } m \geqslant N,$$

where $\mathbb{R}_{00}^{\mathbb{N}} \ni x^m = (x_1, x_2, \ldots, x_m, 0, 0, \ldots)$. It follows that $x^m \to x$ as $m \to \infty$ and density follows.

2 To see this, take any convergent sequence of terms selected from the original sequence and show that its limit must be a term of the sequence also.

Answer to Q8

Since the power set 2^X of any set contains all of its subsets, the conditions **(t1)**–**(t3)** are satisfied because set operations on subsets of X can only produce subsets of X. In this topology every set is open, so that all singletons $\{x\}$ are open. If we have a sequence $(x_n)_{n\in\mathbb{N}}$ that converges to a limit x_∞, it would mean that the open set $\{x_\infty\}$ would have to contain all but finitely many terms of the sequence. This means that the sequences becomes stationary beyond some index $N \in \mathbb{N}$, i. e., that

$$\exists N \in \mathbb{N} \text{ s. t. } x_n = x_\infty \quad \text{for } n \geqslant N.$$

At the other extreme, take any sequence $(x_n)_{n\in\mathbb{N}}$ in X and any $x \in X$, then the only open set containing x is the whole set X since $\tau = \{X, \emptyset\}$, which contains every term of the sequence. Thus any sequence is convergent to any point in X.

Answer to Q9

It clearly holds that $Y = X \cap Y$ and $\emptyset = \emptyset \cap Y$, and hence that $Y, \emptyset \in \tau_{X,Y}$ since $X, \emptyset \in \tau_X$. If $\{U_\lambda \mid \lambda \in \Lambda\}$ is any family of open sets in $\tau_{X,Y}$, then there are open sets $O_\lambda \in \tau_X$ such that $U_\lambda = O_\lambda \cap Y$ and then

$$\bigcup_{\lambda\in\Lambda} U_\lambda = \bigcup_{\lambda\in\Lambda} (O_\lambda \cap Y) = \Big(\bigcup_{\lambda\in\Lambda} O_\lambda\Big) \cap Y,$$

which shows that $\bigcup_{\lambda\in\Lambda} U_\lambda \in \tau_{X,Y}$ since $\bigcup_{\lambda\in\Lambda} O_\lambda \in \tau_X$ as τ_X is a topology. Similarly, given finitely many open sets $U_k \in \tau_{X,Y}$ $(k = 1, \ldots, m)$, there are $O_k \in \tau_X$ with $U_k = O_k \cap Y$, and thus

$$\bigcap_{k=1}^{n} U_k = \bigcap_{k=1}^{n} (O_k \cap Y) = \Big(\bigcap_{k=1}^{n} O_k\Big) \cap Y,$$

showing that $\bigcap_{k=1}^{n} U_k \in \tau_{X,Y}$ since $\bigcap_{k=1}^{n} O_k \in \tau_X$ as τ_X is a topology.

11.5 Answers to Chapter 5 questions

Answer to Q1

Linearity gives that

$$\sup_{0\neq x\in\mathbb{R}^n} \frac{|Mx|_2}{|x|_2} = \sup_{0\neq x\in\mathbb{R}^n} \Big|M\frac{x}{|x|_2}\Big|_2 = \sup_{|y|_2=1} |My|_2,$$

since $|\frac{x}{|x|_2}|_2 = 1$. As for the norm axioms, it clearly holds that $|M|_2 \geqslant 0$, and if $\sup_{|y|_2} |My|_2 = 0$, then it must hold that $My = 0$ for every y of unit norm, and consequently, that $Mx = |x|_2 M \frac{x}{|x|_2} = 0$ for every $x \neq 0$. It follows that $M = 0$ iff $\|M\| = 0$. Next, let $\lambda \in \mathbb{R}$. Then

$$\|\lambda M\| = \sup_{|y|_2=1} |\lambda My|_2 = |\lambda| \sup_{|y|_2=1} |My|_2 = |\lambda|\|M\|,$$

since $|\cdot|_2$ is a norm on $\mathbb{R}^n$. Finally, let $N \in \mathbb{R}^{n\times n}$ notice that

$$\big|(M+N)y\big|_2 \leqslant |My|_2 + |Ny|_2 \quad \text{for } y \text{ with } |y|_2 = 1,$$

and take the supremum over y of unit norm to see that

$$\begin{aligned}\|M+N\| &= \sup_{|y|_2=1}\big|(M+N)y\big|_2 \leqslant \sup_{|y|_2=1}\big[|My|_2 + |Ny|_2\big] \\ &\leqslant \sup_{|y|_2=1} |My|_2 + \sup_{|y|_2=1} |Ny|_2 = \|M\| + \|N\|,\end{aligned}$$

which is the triangle inequality. As for a geometric interpretation of this norm, it can be thought of as the maximal stretching (or shrinking) factor of the linear map. It measures how much the norm of a vector is changed, at most, by applying M.

Answer to Q2

We only show the claim for $BC(\Omega, \mathbb{R}^n)$. The proof has two steps: first, we show that there is a pointwise limit to any Cauchy sequence, then we show that the limit is bounded and continuous. Given a Cauchy sequence $(u_k)_{k\in\mathbb{N}}$, we see that

$$\big|u_k(x) - u_l(x)\big| \leqslant \|u_k - u_l\|_\infty, \quad x \in \Omega,$$

so that, for every $x \in \Omega$, $(u_k(x))_{k\in\mathbb{N}}$ is a Cauchy sequence in $\mathbb{R}^n$ and, as such, has a limit $u_\infty(x)$. We therefore have a candidate for the limiting function. As Cauchy sequences are bounded, we find $M > 0$ such that

$$\big|u_k(x)\big| \leqslant \|u_k\|_\infty \leqslant M, \quad k \in \mathbb{N},\ x \in \Omega,$$

which letting k tend to infinity shows that $|u_\infty(x)| \leqslant M$ for every $x \in \Omega$. Boundedness is thus established. Next, we show that convergence actually occurs in the supremum norm. Given any $\varepsilon > 0$, $K \in \mathbb{N}$ can be found with

$$\big|u_k(x) - u_l(x)\big| \leqslant \|u_k - u_l\|_\infty \leqslant \varepsilon \quad \text{for } k, l \geqslant K \text{ and } x \in \Omega.$$

Letting $l \to \infty$, if follows that

$$|u_k(x) - u_\infty(x)| \leqslant \varepsilon \quad \text{for } k \geqslant K \text{ and } x \in \Omega.$$

Taking the supremum over $x \in \Omega$ yields $\|u_k - u_\infty\|_\infty \leqslant \varepsilon$ for $k \geqslant K$, which gives uniform convergence. Finally, we show that u_∞ is also continuous. Given $x \in \Omega$ and $\varepsilon > 0$, we first choose $K \in \mathbb{N}$ so that $\|u_K - u_\infty\|_\infty \leqslant \frac{\varepsilon}{3}$ and then exploit the continuity of u_K to find $\delta > 0$ such that

$$|u_K(y) - u_K(x)| \leqslant \frac{\varepsilon}{3} \quad \text{for } y \in \mathbb{B}(x, \delta).$$

This gives

$$\begin{aligned} |u_\infty(y) - u_\infty(x)| &\leqslant |u_\infty(y) - u_K(y)| + |u_K(y) - u_K(x)| + |u_K(x) - u_\infty(x)| \\ &\leqslant \varepsilon \quad \text{for } y \in \mathbb{B}(x, \delta), \end{aligned}$$

and the desired continuity follows since $x \in \Omega$ was arbitrary.

Answer to Q3

Let C be a closed subset of the complete metric space M. If $(x_n)_{n\in\mathbb{N}}$ is a Cauchy sequence in C, then it is one in M and, as such, will possess a limit $x_\infty \in M$. Since $x_n \to x_\infty$ as $n \to \infty$, $x_\infty \in C$ since C is closed (and closed sets contain all their limit points).

11.6 Answers to Chapter 6 questions

Answer to Q1

If $M \subset \mathbb{R}^n$ is locally the graph of a function C^1-function f of k-variables, then given $P \in M$, there are a coordinate system $X = (x^1, \ldots, x^n)$ (a point, the origin, which can be taken to be P, and n linearly independent vectors, a basis), an open set $O_P \subset M$ containing P and $f_P \in C^1(\mathbb{B}_{\mathbb{R}^k}(0,1), \mathbb{R}^{n-k})$ such that

$$M \cap O_P = \{(x', f_P(x')) \mid x' \in \mathbb{B}_{\mathbb{R}^k}(0,1)\} \quad \text{and with } f_P(0) = 0,$$

where $x' = (x^1, \ldots, x^k)$. It is left as an exercise to verify that the map

$$\varphi_P : O_P \to \mathbb{R}^k, Q \mapsto x'(Q),$$

has all the required properties, i. e., that φ_P is a homeomorphism onto its range, with differentiable inverse φ_P^{-1} such that $D\varphi_P^{-1}(0)$ is of maximal rank. Notice that it holds that $\varphi_P^{-1}(x') = (x', f_P(x'))$! You are encouraged to draw a picture of the above set up, even simply in the case of a 1-dimensional submanifold of $\mathbb{R}^2$.

Answer to Q2

Indeed, take any open set $O \subset M$. Then, for each $x \in O$ and for each $n \in \mathbb{N}$, we can find $k_n(x) \in \{1, \ldots, N_n\}$ such that $x \in \mathbb{B}(P^n_{k_n(x)}, 1/n)$. Take $n = n(x)$ large enough to ensure that

$$d(x, O^c) \geqslant \frac{2}{n},$$

and observe that

$$d(P^n_{k(x)}, y) \geqslant d(x, y) - d(x, P^n_{k(x)}) \quad \forall y \in O^c,$$

by the triangle inequality. By taking the infimum over $y \in O^c$, we arrive at

$$d(P^n_{k(x)}, O^c) \geqslant d(x, O^c) - d(x, P^n_{k(x)}) \geqslant \frac{2}{n} - \frac{1}{n} = \frac{1}{n},$$

which means that $\mathbb{B}(P^n_{k(x)}, 1/n(x)) \subset O$. It only remains to note that

$$O = \bigcup_{x \in O} \{x\} \subset \bigcup_{x \in O} \mathbb{B}(P^n_{k(x)}, 1/n(x)) \subset O,$$

where the union on the right consists of at most countably many distinct balls.

Answer to Q3

Assume first that $\dot{\gamma}_1(0) = \dot{\gamma}_2(0)$ and take $f \in C^1(\mathbb{R}^n \mathbb{R})$. Then

$$\frac{d}{dt}(f \circ \gamma_1)(0) = Df(P)\dot{\gamma}_1(0) = Df(P)\dot{\gamma}_2(0) = \frac{d}{dt}(f \circ \gamma_2)(0),$$

since $\gamma_1(0) = P = \gamma_2(0)$. Assume now that the identity holds for all differentiable functions $f : \mathbb{R}^n \to \mathbb{R}$. Choosing $f = X^i$, $i = 1, \ldots, n$, where $X^i(x) = x^i$ for $x \in \mathbb{R}^n$, we obtain that

$$\dot{\gamma}_1^i(0) = \frac{d}{dt}(X^i \circ \gamma_1)(0) = \frac{d}{dt}(X^i \circ \gamma_2)(0) = \dot{\gamma}_2^i(0) \quad \text{for } i = 1, \ldots, n,$$

and thus $\dot{\gamma}_1(0) = \dot{\gamma}_2(0)$.

11.7 Answers to Chapter 7 questions

Answer to Q1

Take a Cauchy sequence of functions $(x_n)_{n\in\mathbb{N}}$ in X_0 with respect to $\|\cdot\|_\infty$. Then, given $\varepsilon > 0$, $N \in \mathbb{N}$ can be found such that

$$|x_n(t) - x_m(t)| \leqslant \|x_n - x_m\|_\infty \leqslant \varepsilon \quad \text{for } t \in [0,T] \text{ and } m,n \geqslant N.$$

Thus, for any fixed $t \in [0,T]$, $(x_n(t))_{n\in\mathbb{N}}$ is a Cauchy sequence of real numbers and it necessarily possesses a limit, which we denote by $x_\infty(t)$. Letting m tend to infinity in the above inequality, we see that

$$|x_n(t) - x_\infty(t)| \leqslant \varepsilon \quad \text{for } t \in [0,T] \text{ and } n \geqslant N.$$

Taking the supremum over $t \in [0,T]$, we arrive at

$$\|x_n - x_\infty\|_\infty \leqslant \varepsilon \quad \text{for } n \geqslant N,$$

which yields the convergence of $(x_n)_{n\in\mathbb{N}}$. Observe that $x_\infty(0) = x_0$, since $x_n(0) = x_0$ for every $n \in \mathbb{N}$, and that x_∞ is continuous as the uniform limit of continuous functions. We conclude that $x_\infty \in X_0$ and that X_0 is complete.

Answer to Q2

Notice that the triangle inequality implies that

$$\|L_m\| = \|L_m - L_n + L_n\| \leqslant \|L_m - L_n\| + \|L_n\|.$$

This gives

$$\|L_m\| - \|L_n\| \leqslant \|L_m - L_n\|,$$

and, by switching the roles of m and n, also that

$$\|L_n\| - \|L_m\| \leqslant \|L_n - L_m\| = \|L_m - L_n\|.$$

Combining these two inequalities yields the claim.

Answer to Q3

Assume that $y : \mathbb{R} \to \mathbb{R}^n$ is another solution and consider the function u defined by $u(t) = e^{-tA}y(t)$ for $t \in \mathbb{R}$. Taking a derivative, we see that

$$\dot{u}(t) = -Ae^{-tA}y(t) + e^{-tA}\dot{y}(t) = e^{-tA}(-Ay(t) + Ay(t)) = 0, \quad t \in \mathbb{R}.$$

It follows that $u(t) = u(0) = y(0) = x_0$ for $t \in \mathbb{R}$, which yields $y(t) = e^{tA}x_0$ and the desired uniqueness.

Answer to Q4

Coercivity of u yields that, for any given $M_1 > 0$, $u(z) \geqslant M_1$ provided $|z|_2 \geqslant R$ for some $R > 0$. Continuity of u now guaranties that $u|_{\overline{\mathbb{B}}(0,R)}$ attains a minimum $M_2 \in \mathbb{R}$, i. e., that $u(z) \geqslant M_2$ for $z \in \overline{\mathbb{B}}(0,R)$. The claim follows with $M = \min\{M_1, M_2\}$.

Answer to Q5

Choosing local coordinates $Y : \mathcal{U} \to \mathbb{B}_{\mathbb{R}^m}(0,1)$ about the point P, a solution of the ordinary differential equation is a curve $y : (-\varepsilon,\varepsilon) \to \mathcal{U}$, the coordinates $\alpha = Y \circ y : (-\varepsilon,\varepsilon) \to \mathbb{R}^m$ of which satisfy

$$\begin{cases} \sum_{j=1}^{m} \dot{\alpha}^j \frac{\partial}{\partial y_j} = \sum_{j=1}^{m} \langle dy^j, (F \circ Y^{-1})(\alpha)\rangle \frac{\partial}{\partial y_j}, \\ \alpha(0) = Y(P) = 0. \end{cases}$$

11.8 Answers to Chapter 8 questions

Answer to Q2

Let us first compute the subdifferential in $x = 0$. Taking $v \in \overline{\mathbb{B}}_{|\cdot|_\infty}(0,1)$, we see that

$$v \cdot h + |0|_1 = v_1h_1 + \cdots + v_dh_d \leqslant |v_1||h_1| + \cdots + |v_d||h_d| \leqslant |0 + h|_1,$$

since $|v_i| \leqslant 1$ for $i = 1, \ldots, d$. Conversely, if it holds that

$$|h_1| + \cdots + |h_n| \geqslant v \cdot h = v_1h_1 + \cdots + v_dh_d$$

for every $h \in \mathbb{R}^d$, we can conclude that

$$|v_i| \leqslant 1 \quad \text{for } i = 1, \ldots, d,$$

by setting $h = \operatorname{sign}(v_i)e_i$ for $i = 1, \ldots, d$. This means that

$$\partial|\cdot|_1(0) = \mathbb{B}_{|\cdot|_\infty}(0,1).$$

As for the connection with $\{(\mathrm{sign}(x_1), \ldots, \mathrm{sign}(x_d))\}$, notice that approaching the origin from different directions x, one can generate any vector of the form $(\pm 1, \ldots, \pm 1)$. These vectors are the vertices (extremal points) of $\mathbb{B}_{|\cdot|_\infty}(0,1) = \partial|\cdot|_1(0)$.

Answer to Q1

If we assume that the matrix A has positive eigenvalues, then

$$A = \sum_{j=1}^{n} \lambda_j u_j u_j^\top \quad \text{and} \quad \lambda_j > 0 \quad \text{for } j = 1, \ldots, n$$

and

$$x^\top A x = x^\top \sum_{j=1}^{n} \lambda_j u_j u_j^\top x = \sum_{j=1}^{n} \lambda_j (u_j^\top x)^2 = \sum_{j=1}^{n} \lambda_j (\beta^j)^2,$$

if $x = \sum_{j=1}^{n} \beta^j u_j$ (i. e., using the eigenvectors as a basis). It follows that

$$\{x \in \mathbb{R}^n \mid x^\top A x = 1\} = \left\{\beta \in \mathbb{R}^n \,\middle|\, \sum_{j=1}^{n} \lambda_j (\beta^j)^2 = 1\right\},$$

and that the set is indeed an ellipse. We also recognize that the eigenvectors point in the direction of the main axes of the ellipse, which have length $\frac{2}{\sqrt{\lambda_j}}$ for $j = 1, \ldots, n$.

11.9 Answers to Chapter 9 questions

Answer to Q1

Since we already know that $\ell_2(\mathbb{Z}^d, \mathbb{C})$ is complete, it is enough to show that $c_{00}(\mathbb{Z}^d, \mathbb{C})$ is dense in it. For that, take $\hat{u} = (\hat{u}_k)_{k \in \mathbb{N}} \in \ell_2(\mathbb{Z}^d, \mathbb{C})$, and for any $n \in \mathbb{N}$, set

$$\hat{u}^n = (\hat{u}_1, \ldots, \hat{u}_n, 0, 0, \ldots) \in c_{00}(\mathbb{Z}^d, \mathbb{C}).$$

It remains to observe that

$$\|\hat{u} - \hat{u}^n\|_{\ell_2}^2 = \sum_{k=n+1}^{\infty} \hat{u}_k^2 \longrightarrow 0 \quad \text{as } n \to \infty,$$

since $\hat{u} \in \ell_2(\mathbb{Z}^d, \mathbb{C})$. To see that density is enough, take a Cauchy sequence $\hat{u}^n$ in $c_{00}(\mathbb{Z}^d, \mathbb{C})$ with respect to $\|\cdot\|_{\ell_2}$. As $c_{00}(\mathbb{Z}^d, \mathbb{C}) \subset \ell_2(\mathbb{Z}^d, \mathbb{C})$ and the latter is complete, the sequence will have a limit $\hat{u} \in \ell_2(\mathbb{Z}^d, \mathbb{C})$ with respect to $\|\cdot\|_{\ell_2}$. This shows that the completion $\overline{c_{00}(\mathbb{Z}^d, \mathbb{C})}$ satisfies the inclusion

$$\overline{c_{00}(\mathbb{Z}^d, \mathbb{C})} \subset \ell_2(\mathbb{Z}^d, \mathbb{C}),$$

and density that they actually coincide.

Answer to Q2

It is a calculation to see that

$$\hat{h}_k = (h|\varphi_k) = \frac{1}{\sqrt{2\pi}} \int_0^{2\pi} e^{-ikx} h(x)\, dx = \sqrt{\frac{2}{\pi}} \frac{i}{k} [1 + (-1)^{k+1}], \quad k \in \mathbb{Z} \setminus \{0\}.$$

Since $|\hat{h}_k|^2 = \frac{c}{k^2}$, $k \neq 0$, it follows that $\hat{h} \in \ell_2(\mathbb{Z}, \mathbb{C})$, i. e., that $h \in \mathrm{L}^2(\mathbb{T})$. Clearly, h is not periodic as $h(0) \neq h(2\pi)$.

Answer to Q3

Taking $u \in \mathrm{H}^s_\pi$ means that

$$\sum_{k \in \mathbb{Z}^d} [1 + |k|_2^2]^s |\hat{u}_k|^2 < \infty.$$

Observing that $-\widehat{(\Delta u)}_k = |k|_2^2 \hat{u}_k$, it will not be true in general that

$$\| - \Delta u\|_{s,2}^2 = \sum_{k \in \mathbb{Z}^d} [1 + |k|_2^2]^s |k|_2^4 \, |\hat{u}_k|^2 < \infty.$$

Nevertheless, one has that

$$\begin{aligned} \| - \Delta u\|_{s-2,2}^2 &= \sum_{k \in \mathbb{Z}^d} [1 + |k|_2^2]^{s-2} |k|_2^4 \, |\hat{u}_k|^2 \\ &\leqslant \sum_{k \in \mathbb{Z}^d} [1 + |k|_2^2]^{s-2} [1 + |k|_2^2]^2 \, |\hat{u}_k|^2 = \|u\|_{s,2}. \end{aligned}$$

Answer to Q4

Since $u \in \mathcal{D}$, u has compact support and

$$e^{-ix\cdot\xi} u(x) = 0 \quad \text{for } |x| \geqslant M \text{ and some } M > 0.$$

As a continuous function, u is bounded on its support. Then

$$\begin{aligned}|e^{-ix\cdot\xi}u(x) - e^{-ix\cdot\xi_n}u(x)| &\leqslant C|e^{-ix\cdot\xi} - e^{-ix\cdot\xi_n}|\\ &= C\left|\int_0^1 e^{-ix\cdot(\xi_n+\tau(\xi-\xi_n))}x\cdot(\xi-\xi_n)\,d\tau\right|\\ &\leqslant CM|\xi-\xi_n|,\end{aligned}$$

which yields the stated convergence uniformly in x. If you do not see the latter, carry out the missing details showing that the integrals defining $\hat{u}$ do indeed converge.

11.10 Answers to Chapter 10 questions

Answer to Q1

Take a sequence $(E_n)_{n\in\mathbb{N}}$ of pairwise disjoint events. Then, defining

$$F_n = \bigcup_{k=1}^{n} E_n \quad \text{for } n \in \mathbb{N},$$

one obtains an increasing sequence of sets and, therefore,

$$P\left(\bigcup_{n=1}^{\infty} F_n\right) = \lim_{n\to\infty} P\left(\bigcup_{k=1}^{n} F_k\right)$$

by assumption. It holds that

$$\bigcup_{k=1}^{n} F_k = \bigcup_{k=1}^{n} E_k$$

and, therefore, that

$$P\left(\bigcup_{k=1}^{n} F_k\right) = P\left(\bigcup_{k=1}^{n} E_k\right) = \sum_{k=1}^{n} P(E_k),$$

by (finite) additivity. We conclude that

$$\begin{aligned}P\left(\bigcup_{n=1}^{\infty} E_n\right) = P\left(\bigcup_{n=1}^{\infty} F_n\right) &= \lim_{n\to\infty} P\left(\bigcup_{k=1}^{n} F_k\right) = \lim_{n\to\infty} P\left(\bigcup_{k=1}^{n} E_k\right)\\ &= \lim_{n\to\infty}\sum_{k=1}^{n} P(E_k) = \sum_{k=1}^{\infty} P(E_k),\end{aligned}$$

which is σ-additivity.

Answer to Q2

Take the set

$$E_n = \left\{ x \in \mathbb{R} \;\middle|\; P(X = x) \geqslant \frac{1}{n} \right\}$$

and observe that it contains at most n elements since otherwise its probability would exceed 1. In particular, the sets E_n are finite and, therefore, their union is at most countable. Since

$$\{x \in \mathbb{R} \mid P(X = x) > 0\} = \bigcup_{n \in \mathbb{N}} E_n,$$

the claim follows.

Answer to Q3

Monotonicity follows observing that

$$\{X \leqslant x\} \subset \{X \leqslant \tilde{x}\} \quad \text{for } x \leqslant \tilde{x},$$

since $P(E) \leqslant P(F)$ whenever $E \subset F$. In order to verify the first property, observe that

$$\bigcap_{n \in \mathbb{N}} \{X \leqslant -n\} = \emptyset$$

and that $\{X \leqslant -n - 1\} \subset \{X \leqslant -n\}$ for $n \in \mathbb{N}$. The continuity property of probability therefore yields that

$$0 = P(\emptyset) = P\left(\bigcap_{n \in \mathbb{N}} \{X \leqslant -n\} \right) = \lim_{n \to \infty} P(X \leqslant -n) = \lim_{n \to -\infty} F(n).$$

Monotonicity ensures that n can be replaced by x in the limit since it can be verified that $\{X \leqslant x\} \subset \{X \leqslant -n\}$ for $x \leqslant -n$. The second identity follows in a similar way considering the increasing sequence of sets $\{X \leqslant n\}$ for $n \in \mathbb{N}$. As for right continuity, consider

$$\{X \leqslant x\} = \bigcap_{n \in \mathbb{N}} \left\{ X \leqslant x + \frac{1}{n} \right\},$$

and exploit continuity of probability and monotonicity of F. The existence of left limits follows from monotonicity (why?). Why does left continuity not hold in general? Give an example.

Answer to Q4

Let $w \in \mathbb{R}^m$ be a unit vector along which we would like to project the data so that it has maximal variance. In other words, we would like to choose w so that the scalar data set

$$w^\top x_1 = (\mathbb{X}w)^1, \dots, w^\top x_n = (\mathbb{X}w)^n$$

has maximal sample variance $\sum_{i=1}^n (w^\top x_i)^2 = \sum_{i=1}^n [(\mathbb{X}w)^i]^2$. Notice that the mean of the projected data vanishes since

$$w^\top x_1 + \cdots + w^\top x_n = w^\top (x_1 + \cdots + x_n) = w^\top 0 = 0,$$

as the original data is centered. Using that

$$\mathbb{X} = \sum_{j=1}^K \sigma_j v_j u_j^\top,$$

the variance computes to

$$\begin{aligned}
\sum_{i=1}^n \left(\sum_{j=1}^K \sigma_j v_j^i (u_j^\top w) \right)^2 &= \sum_{i=1}^n \sum_{j,l=1}^K \sigma_j \sigma_l v_j^i v_l^i (u_j^\top w)(u_l^\top w) \\
&= \sum_{j,l=1}^K \sigma_j \sigma_l (u_j^\top w)(u_l^\top w) \left(\sum_{i=1}^n v_j^i v_l^i \right) \\
&= \sum_{j=1}^K \sigma_j^2 (u_j^\top w)^2,
\end{aligned}$$

since $v_1, \dots, v_K$ is an orthonormal system. As $u_1, \dots, u_K$ also form an orthonormal system, the maximum is achieved by choosing $w = u_1$ as σ_1 is the largest singular value. The next best choice is the direction u_2 associated to the next largest singular value σ_2. We allow singular values to repeat if necessary and "next best" has to be interpreted accordingly.

Index

https://doi.org/10.1515/9783110780925-012

www.ingramcontent.com/pod-product-compliance
Lightning Source LLC
Chambersburg PA
CBHW081103220326
41598CB00038B/7210
* 9 7 8 3 1 1 0 7 8 0 8 5 7 *